# FAULT DIAGNOSIS & PROGNOSIS TECHNIQUES FOR COMPLEX ENGINEERING SYSTEMS

# FAULT DIAGNOSIS & PROGNOSIS TECHNIQUES FOR COMPLEX ENGINEERING SYSTEMS

BY

GHY GRANT
CHARLOTTE MURPHY

KNOWLEDGE
BAKERS

KNOWLEDGE BAKERS

ISBN: 978-93-9001-385-2 (HB)



*Published by* : **Knowledge Bakers**
Pune, India
*Email: info@knowledgebakers.com*

# PREFACE

In the realm of engineering, the reliability and performance of complex systems are paramount for ensuring safety, efficiency, and sustainability in various industrial sectors, including aerospace, automotive, energy, manufacturing, and more. However, the inherent complexity and interconnectedness of modern engineering systems pose significant challenges in detecting, diagnosing, and predicting faults that may compromise their functionality and integrity. As engineers strive to enhance the reliability and maintainability of these systems, fault diagnosis and prognosis emerge as critical disciplines that play a vital role in minimizing downtime, optimizing maintenance strategies, and prolonging the operational lifespan of critical assets.

This book, titled "Fault Diagnosis and Prognosis Techniques for Complex Engineering Systems," aims to provide a comprehensive overview of the principles, methodologies, and cutting-edge techniques employed in the diagnosis and prognosis of faults in complex engineering systems. Through collaboration with leading experts and practitioners in the field, we endeavor to explore the latest advancements, challenges, and opportunities in fault diagnosis and prognosis, spanning both theoretical developments and practical applications across diverse industries.

As editors of this volume, we have curated a collection of chapters that cover a wide range of topics within fault diagnosis and prognosis, including signal processing techniques, machine learning algorithms, statistical analysis methods, data-driven modeling approaches, and case studies in various engineering domains. Each chapter offers insights from both theoretical and practical perspectives, drawing on real-world examples, industry best practices, and interdisciplinary connections to showcase the multifaceted nature of fault diagnosis and prognosis in complex engineering systems.

Furthermore, this book seeks to highlight the importance of proactive maintenance strategies and predictive analytics in optimizing the performance and reliability of engineering systems. By exploring the synergies between fault diagnosis, prognosis, and condition-based maintenance, we aim to empower engineers and practitioners to adopt data-driven approaches and advanced techniques for anticipating and mitigating potential failures, thereby enhancing operational efficiency and minimizing risks in critical infrastructure and industrial processes.

As we embark on this enlightening journey through the realm of fault diagnosis and prognosis, we extend our gratitude to the contributors whose expertise and dedication have enriched this volume. We hope that "Fault Diagnosis and Prognosis Techniques for Complex Engineering Systems" serves as a valuable resource for researchers, engineers, students, and practitioners alike, fostering collaboration, innovation, and progress in the ever-evolving field of engineering reliability and maintenance.

**Ghy Grant**
**Charlotte Murphy**

# CONTENTS

Chapter 1

# Introductory Chapter: Prognostics - An Overview

## 1. Introduction

Prognostics, in general, can be defined as "knowledge beforehand". Prognostics is usually identified with medical issues. Nowadays, due to the new advances in technologies and information systems, prognostic is beginning to be employed in other fields, e.g. engineering, financial, business, etc.

The main key indicators are given by European Standard EN 15341:2007 [1]. The objectives of the key indicators are to measure the status, compare (internal and external benchmarks), diagnose (analysis of strengths and weaknesses), identify objectives and define targets to be reached, plan improvement actions and continuously measure changes over time. There are three main groups of indicators: economic [2], technical [3] and organisational [4]. They are set considering endogenous (company culture, industry, life cycle of the components, criticality, etc.) and exogenous (location, society culture, market, laws, regulations, etc.) variables [5].

Prognostics requires also of advance analytics in order to format, save and analyse these signals and information, from qualitative and quantitative points of view. *Model-based approach* takes into account the state prediction achieved through physics or system models, the following being mainly employed: model based on detection and isolation [6]. Hybrid models, extended Kalman filtering and particle filtering [7, 8]. *Data-driven approach* is also a state prediction with criteria evaluation, where the state prediction is achieved through regression or stochastic process modelling. The most important are autoregressive moving average (ARMA) or autoregressive integrated moving average (ARIMA) [9], etc.

Nowadays the information from an item or person is getting more and more, with more variables, complex, etc. The large amount of data requires to be analysed, considering the heterogeneity, noise accumulation, spurious correlations, and incidental endogeneity of the data. It does that new approach and algorithms based on artificial intelligence which will be appearing; Artificial Neural Network [10]; Fuzzy Logic System [11]; Hidden Markov Model [12]; Support Vector Machine [13], Relevance Vector Machine (RVM); Gaussian Process Regression [14], Multivariate Logistic Regression in general form, K-Means Clustering, Fuzzy Logic-Based Algorithms and Bayesian Belief Network, etc. Some algorithms can be applied together with the above-mentioned methods, e.g. gradient descent, alternating least squares, collaborative filtering, SVM kernel, belief propagation, matrix factorization and Gibbs sampling.

The next generation of approaches will require to process Big Data. Big Data is one of the central and influential research challenges for the 2020 Horizon, where the quantity of world data will be 44 times bigger in the next few years (0.8–35 zettabytes) [15].

## References

[1] EN_15341:2007. Maintenance—Maintenance Key Performance Indicators. Europeam Standard; 2010

[2] Pérez JMP, Asensio ES, Márquez FPG. Economic viability analytics for wind energy maintenance management. In: Advanced Business Analytics. Springer; 2015. pp. 39-54

[3] Jiménez AA, Muñoz CQG, Márquez FPG. Machine learning and neural network for maintenance management. In: International Conference on Management Science and Engineering Management. Springer; 2017. pp. 1377-1388

[4] Marugán AP, Márquez FPG. Improving the efficiency on decision making process via BDD. In: Proceedings of the Ninth International Conference on Management Science and Engineering Management. Springer; 2015. pp. 1395-1405

[5] Pliego Marugán A, García Márquez FP, Lev B. Optimal decision-making via binary decision diagrams for investments under a risky environment. International Journal of Production Research. 2017;**55**:5271-5286

[6] Ray A, Tangirala S. Stochastic modeling of fatigue crack dynamics for on-line failure prognostics. IEEE Transactions on Control Systems Technology. 1996;**4**:443-451

[7] Márquez FPG, Zaman N. Digital Filters and Signal Processing. Intech; 2013

[8] Márquez FPG. Digital Filters. Intech; 2011

[9] Ho S, Xie M. The use of Arima models for reliability forecasting and analysis. Computers and Industrial Engineering. 1998;**35**:213-216

[10] Marugán AP, Márquez FPG, Perez JMP, Ruiz-Hernández D. A survey of artificial neural network in wind energy systems. Applied Energy. 2018;**228**:1822-1836

[11] Benmessaoud T, Marugán AP, Mohammedi K, Márquez FPG. Fuzzy logic applied to scada systems. In: International Conference on Management Science and Engineering Management. Springer; 2017. pp. 749-757

[12] Fine S, Singer Y, Tishby N. The hierarchical hidden Markov model: Analysis and applications. Machine Learning. 1998;**32**:41-62

[13] Manupati V, Anand R, Thakkar J, Benyoucef L, Garsia FP, Tiwari M. Adaptive production control system for a flexible manufacturing cell using support vector machine-based approach. The International Journal of Advanced Manufacturing Technology. 2013;**67**:969-981

[14] Quiñonero-Candela J, Rasmussen CE. A unifying view of sparse approximate gaussian process regression. Journal of Machine Learning Research. 2005;**6**:1939-1959

[15] Márquez FPG, Lev B. Big Data Management. Springer; 2017

Chapter 2

# Prognostics 102: Efficient Bayesian-Based Prognostics Algorithm in MATLAB


## Abstract

An efficient Bayesian-based algorithm is presented for physics-based prognostics, which combines a physical model with observed health monitoring data. Unknown model parameters are estimated using the observed data, from which the remaining useful life (RUL) of the system is predicted. This paper focuses on the Bayesian method for parameter estimation of a damage degradation model where epistemic uncertainty in model parameters is reduced with the observed data. Markov-chain Monte Carlo sampling is used to generate samples from the posterior distribution, which are then propagated through the physical model to estimate the distribution of the RUL. A MATLAB script of 76 lines is included in this paper with detailed explanations. A battery degradation model and crack growth model are used to explain the process of parameter estimation, the evolution of degradation and RUL prediction. The code presented in this paper can easily be altered for different applications. This code may help beginners to understand and use Bayesian method-based prognostics.



**Keywords:** Bayesian method, physics-based prognostics, remaining useful life, MATLAB code, crack growth, battery degradation


## 1. Introduction

Structural health monitoring (SHM) [1, 2] is the process of identifying damage and evaluating the safety of a system based on online and/or off-line data. It uses an array of sensors to obtain measurement data that are directly or indirectly related to damage. The statistical analysis of these measurements can help predict the future state of the system and thus improve the safety of the system. SHM can be found in a wide variety of applications such as bridges and dams, buildings, stadiums, platforms, airframes, turbines, etc. Prognostics is an extension of SHM, which is the process of estimating the time beyond which a system can no longer function to meet desired performances [3]. The time, in terms of cycles/hours, remaining to run the system before it fails is called the remaining useful life (RUL).

There are two types of prognostics methods: data-driven and physics-based approaches. The data-driven approaches are advantageous when many training data are available for a complex system, while the physics-based approaches are good when a physical model of damage degradation is available. The physics-based approach is used for prognostics in this paper with a well-defined physics model.

Measured data is used to estimate model parameters, which are then used to predict the RUL.

Recently, many prognostics algorithms have been published in the literature [4–8]. However, many of the proposed algorithms are complex and not easily applicable. This complexity can present a serious hurdle for the beginner. In addition, using commercial programs may not be the best choice in teaching algorithms to students. As a continuation of our educational paper on prognostics algorithm [9], the objective of this paper is to explain the fundamentals of a Bayesian-based prognostics method and demonstrate how to use it using a simple MATLAB code.

The MATLAB code consists of 76 lines, which is further divided into three parts: (1) problem definition; (2) prognostics using the Bayesian method (BM); and (3) post-processing. The program is structured in such a way that the users only need to modify the problem definition part for their own application. This paper shows an example of battery degradation and crack growth models, and attempts to explain prognostics using BM with MATLAB code.

The remaining sections are organized as follows: In Section 2, the overall process of BM is explained; in Section 3, implementation of the code is explained with details using battery degradation example; and in Section 4, modification of the code for crack growth example is described, followed by conclusions in Section 5.

## 2. Methodology

In this section, a physics-based approach is explained using the procedure shown in **Figure 1**. The theoretical discussions in this section are mainly to help understand the MATLAB implementation in Section 3. The physics-based approach comprises of the following steps: (1) developing or identifying a physical model that describes the degradation of system health, (2) collecting data by operating the system under usage conditions and measuring degradation at a sequence of times/cycles, (3) estimating the model parameters by fitting the measured data, (4) progressing the physical model to the future times/cycles, and (5) predicting the RUL. A statistical

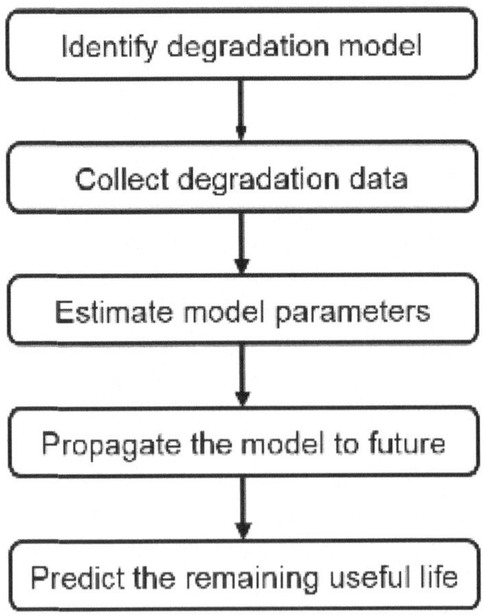


**Figure 1.**
*Flowchart of physics-based prognostics.*

inference technique called the Bayesian method (BM) is used in this paper to estimate the model parameters based on measured data. Many other methods, such as particle filter and Kalman filter, also use Bayesian inference to estimate the model parameters. In BM, all model parameters are estimated in the form of a joint probability density function (PDF), whose distribution can be represented using samples. Among various sampling methods, Markov-chain Monte Carlo (MCMC) algorithm is employed to draw samples from the distribution. These samples of model parameters are then substituted in the physical model to calculate the samples of RUL, from which the statistical distribution is evaluated.

## 2.1 Model definition

In this section, a degradation model of a battery is used to explain the physics-based prognostics algorithm using the Bayesian method. The degradation model of crack growth will also be explained in Section 4. It is expected that the users develop a degradation model for their own application. This section explains the basic requirements of a degradation model.

In a lithium-ion battery, it is well known that the capacity of a secondary cell degrades over cycles in use. Therefore, the capacity can be used as a degradation feature. The degradation feature is an output of the degradation model that shows a monotonic trend as a function of time. The system is considered failed when the degradation feature goes beyond a threshold. In the case of a lithium-ion battery, the failure threshold is defined when the charging capacity fades by 30% of that of a pristine battery. In this paper, the C/1 capacity (capacity at a nominally rated current of 1A) of the battery is used as a degradation feature. Since the C/1 capacity is inversely proportional to the sum of the transfer resistance and the electrolyte resistance, it represents the overall performance of a battery.

Although the degradation process of a battery is complicated, a simple empirical model is available when the usage of the battery is the repetition of fully charging-discharging cycles. In such a case, the degradation model can be written as a function of time only. Since the capacity of a battery degrades over time, the ratio of the capacity compared to that of the pristine battery is expressed by an exponential decaying model as [10]

$$\tilde{y}(t;b) = \exp(-bt) \tag{1}$$

where $b$ is the model parameter, $t$ is the time, and $\tilde{y}(t;b)$ is the relative degradation of the C/1 capacity compared to the pristine capacity. The notation in Eq. (1) is chosen such that the degradation model depends on both the time of usage and model parameters. In general, the degradation model is either monotonically decreased (e.g., the capacity decay of a battery) or monotonically increased (e.g., crack growth).

The main goal of the physics-based prognostics is to predict the degradation behavior using the degradation model. If the model is perfect, then it can be used to find the time $t_{\mathrm{EOL}}$ at the end of life from

$$\tilde{y}(t_{\mathrm{EOL}};b) = y_{\mathrm{threshold}} \tag{2}$$

where $y_{\mathrm{threshold}}$ is the failure threshold. Since the relative capacity is used as a degradation feature, the failure threshold is defined at $y_{\mathrm{threshold}} = 0.7$. Let the current time be $t_{\mathrm{CUR}}$, then the RUL can be defined as

$$t_{\mathrm{RUL}} = t_{\mathrm{EOL}} - t_{\mathrm{CUR}} \tag{3}$$

In practice, however, the degradation model is not perfect in the sense that the model form, as well as the model parameters, may not be accurate. The error in the model form can be handled by introducing a model form error and identifying the error using measured data, which would be considered as the out of the scope of this paper. Interested readers can refer to Guan et al. [11].

Once the model form is accepted, the next task is to identify the accurate model parameters. In the case of the battery degradation model in Eq. (1), the parameter $b$ needs to be identified. In most cases, the model parameters are not an intrinsic property but depend on operating conditions and environment. Therefore, these parameters can be different for different batteries and need to be identified for the specific battery of interest. In fact, the major task of physics-based prognostics is to identify the model parameters.

The model parameter $b$ for a specific battery can be identified by measuring the capacity degradation during regular operation. The measuring process is often called health monitoring, where the degradation feature is measured over time. It is possible that the degradation feature can be monitored online. However, for the purpose of prognostics, the real-time continuous monitoring may not be necessary. Therefore, it is often suggested to collect data in a discrete set of times. Then, many different physics-based prognostics algorithms use these data to identify the model parameters so that the degradation model represents the degradation feature the best. For example, nonlinear least-squares method minimizes the error between measured data and the model prediction. Kalman filter, particle filter, and Bayesian methods are using Bayesian inference to estimate the model parameters. Different methods use different assumptions and different numerical approaches. Interested readers are referred to Kim et al. [3] for details of these methods.

If the measured data are accurate, then a small number of measured data should be good enough to estimate the model parameters. In reality, however, most measured data include noise and bias, which make the estimation process difficult. Noise is a random fluctuation of signals due to uncontrollable factors in the measurement environment, while bias is a systematic departure from the average data. If the measurement is repeated, noise can be changed, while the bias may remain the same. The bias can occur because of calibration error of the sensors, but it may also occur due to the model form error. The effect of the model form error can partially be addressed by introducing the bias in the estimation process. Bias can be added in the model as an extra term and estimated in the same way as other parameters. The distribution of estimated bias is a good indicator if the model can represent degradation data well enough. If estimated bias is widely distributed, it means model form error is large. If it is narrowly distributed and the mean is close to 0, it means the model is accurate. Since noise is random, it is important to compensate for its effect in the parameter estimation process. It is obvious that a large level of noise makes the process difficult. Therefore, it is important to keep the signal-to-noise ratio as high as possible. It is also important to understand the statistical characteristics of the noise. In this paper, it is assumed that the noise follows a Gaussian distribution with a zero mean and unknown standard deviation. On the other hand, the effect of bias will not be considered. Therefore, in addition to the unknown model parameters, it is necessary to estimate the unknown standard deviation of noise in data.

Because of noise and bias, it is often expected that a large number of data be required to estimate the model parameters accurately. In prognostics, it is often assumed that $N_{\text{data}}$ data are collected from a start time to the current time $t \in [t_0, t_{\text{CUR}}]$. Time does not have to be a physical time; it can also be the number of cycles of operation. Then, the model parameters are estimated using $N_{\text{data}}$ data, and future degradation is predicted using the degradation model with the estimated parameters. In particular, the goal is to accurately predict the end of life in Eq. (2) and the RUL in Eq. (3).

It is important to note that the data should show a significant change in the damage feature over time. In the case of crack growth in Section 4, for example, when the crack size is small, it grows very slowly. Therefore, the measurement data in an early stage do not show a significant change in the crack size. In such a case, the signal-to-noise ratio is too low and it is difficult to estimate the model parameters.

In this paper, instead of measuring the degradation of a real battery, the degradation data are generated based on an assumed true model. This has a couple of advantages. First, since the true model and its model parameters are known, it is possible to evaluate the accuracy of the estimation process and that of the RUL. It also allows us to investigate the effect of noise on the performance of prognostics algorithms. In this paper, the relative capacity data are generated based on Eq. (1) with the true model parameter $b_{true} = 0.012$. It is assumed that the C/1 capacity of the battery is measured once a week up to the ninth week. In order to simulate the real measurement environment, a Gaussian noise $\varepsilon \sim N(0, 0.005^2)$ is added to the true data. The following MATLAB commands can be used to generate the measured data:

```
>> time=(0:9)';
>> b=0.012;
>> trueData=exp(-b*time);
>> measuData=trueData+0.005*randn(10,1);
```

Once the measurement data are generated, the true model parameters and the information of noise are not used. **Table 1** and **Figure 2** show the true degradation data and simulated measurement data up to the current time $t_{CUR} = 9$ weeks. Based on the true model, the end of life of the battery is $t_{EOL} = 29.72$ weeks, and thus, the true RUL should be $t_{RUL} = 20.72$ weeks.

## 2.2 Bayesian parameter estimation

Once the measurement data are available, the next step would be to estimate the model parameters. Among many parameter estimation methods, the Bayesian inference is explained in this section. In the following explanation, $\Theta$ represents the random variable of the unknown model parameter, and $Y$ represents the random variable of degradation feature. A variable with an upper case denotes a random variable, while a variable with a lower case denotes a realization of the random variable. Bayesian inference estimates the degree of belief in a hypothesis based on collected evidence. Bayes [12] formulated the degree of belief using the identity in conditional probability:

$$P(\Theta \cap Y) = P(\Theta|Y)P(Y) = P(Y|\Theta)P(\Theta) \quad (4)$$

where $P(\Theta|Y)$ is the conditional probability of $\Theta$ given $Y$. In the case of estimating the model parameter using measured data, the conditional probability of $\Theta$ when the probability of measured data $Y$ is available can be written as

| Time (weeks) | 0 | 1 | 2 | 3 | 4 | 5 | 6 | 7 | 8 | 9 |
|---|---|---|---|---|---|---|---|---|---|---|
| True degradation | 1.000 | 0.988 | 0.976 | 0.965 | 0.953 | 0.942 | 0.931 | 0.919 | 0.909 | 0.898 |
| Measured degradation | 0.995 | 0.983 | 0.975 | 0.974 | 0.942 | 0.938 | 0.930 | 0.920 | 0.911 | 0.895 |

**Table 1.**
*Measurement data (relative capacity) for the battery degradation example.*

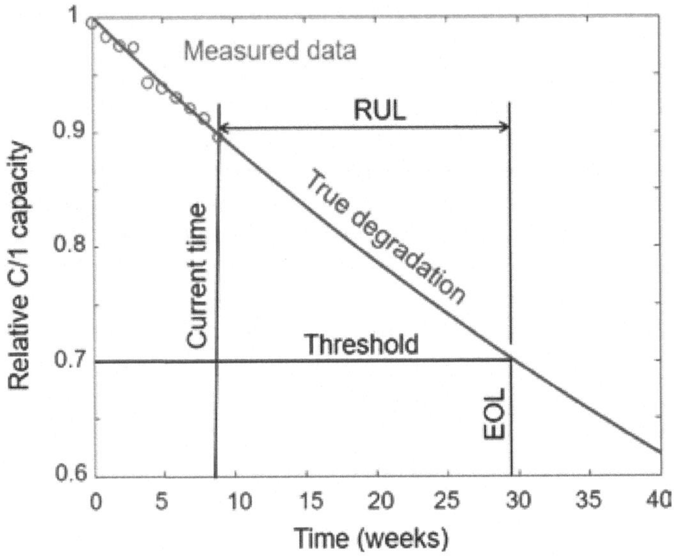


**Figure 2.**
*True degradation curve and measured data for the relative capacity.*

$$P(\Theta|Y) = \frac{P(Y|\Theta)P(\Theta)}{P(Y)} \tag{5}$$

where $P(\Theta)$ is the prior probability of parameter $\Theta$, which represents the preexisting knowledge on the parameter. $P(\Theta|Y)$ is the posterior probability of parameter $\Theta$ after updating the prior with measurement data $Y$. $P(Y|\Theta)$ is the likelihood function or the probability of obtaining data $Y$ for a given parameter $\Theta$. The measurement data affect the posterior probability through the likelihood function. The denominator, $P(Y)$, is the marginal probability of $Y$ and acts as a normalizing constant. The above equation can be used to improve the knowledge of $P(\Theta)$ when additional information $P(Y)$ is available.

If the Bayes' theorem in Eq. (5) is going to be used for identifying unknown model parameters, it would be better to express the theorem in the form of a probability density function (PDF) [13], which is used in the present paper. Let $f_\Theta(\theta)$ be a PDF of model parameter $\Theta$. When there are more than one model parameters, $f_\Theta(\theta)$ can be a joint PDF of multiple parameters. If the health monitoring measures a degradation feature $Y$, the measurement variability can be represented using PDF, $f_Y(y)$. Then, the conditional PDFs between $\Theta$ and $Y$ can be related to the joint PDF and the marginal PDF, $f_\Theta(\theta)$ and $f_Y(y)$, as

$$f_{\Theta Y}(\theta, y) = f_\Theta(\theta|Y = y)f_Y(y) = f_Y(y|\Theta = \theta)f_\Theta(\theta) \tag{6}$$

It is obvious that the joint PDF can be written as $f_{\Theta Y}(\theta, y) = f_\Theta(\theta)f_Y(y)$ when $\Theta$ and $Y$ are independent, and Bayesian inference cannot be used to improve the probability distribution of $f_\Theta(\theta)$. Similar to Eqs. (5) and (6) can be used for obtaining the Bayesian inference in the form of PDF as [14].

$$f_\Theta(\theta|Y = y) = \frac{f_Y(y|\Theta = \theta)f_\Theta(\theta)}{f_Y(y)} \tag{7}$$

Since the denominator $f_Y(y)$ is a constant and since the integral of $f_\Theta(\theta|Y = y)$ is one from the property of PDF, the denominator in Eq. (7) can be considered as a normalizing constant. Similar to Eq. (5), $f_\Theta(\theta|Y=y)$ is the posterior PDF of parameter $\Theta$ that is updated from the prior PDF $f_\Theta(\theta)$ with the likelihood function $f_Y(y|\Theta=\theta)$, which is the probability density value of measured data $y$ given model parameter $\Theta=\theta$. The process of updating the posterior distribution $f_\Theta(\theta|Y=y)$ of the model parameter using the measured data $y$ is called Bayesian inference.

The Bayesian inference can be extended to the case when many data are available. In general, it is possible that the posterior PDF can be obtained by applying all data simultaneously or by iteratively applying each data at a time. Although two approaches are theoretically equivalent, they end up numerically different methods. For example, the particle filter method uses a single measurement to update the posterior distribution, and the previous posterior distribution is used as a prior distribution for the following measurement. On the other hand, Bayesian method uses all measurement data together to build a single posterior distribution, which is used in this paper. Let us consider that $\mathbf{y}=\left\{y_1, y_2, \ldots, y_{N_{\text{data}}}\right\}$ is the vector or $N_{\text{data}}$ measurements. In such a case, the Bayes' theorem can be written as

$$f_\Theta(\theta|Y=\mathbf{y}) = \frac{1}{K}\prod_{i=1}^{N_{\text{data}}}\left[f_Y(y_i|\Theta=\theta)\right]f_\Theta(\theta) \tag{8}$$

where $K$ is the product of all marginal PDFs. However, it can be considered as a normalizing constant to make the integration of the posterior PDF to be one. It is noted that the total likelihood function is the product of the likelihood functions of individual data, which is then multiplied by the prior PDF followed by normalization to yield the posterior PDF.

In contrast to the traditional least-squares method, the Bayes' theorem can estimate not only the best values of parameters but also the uncertainty structure of the identified parameters. Since these uncertainty structures are derived from that of the prior distribution and likelihood function, the uncertainty of the posterior distribution is directly related to that of the likelihood and the prior distribution.

In the Bayesian method, it is assumed that the users know the prior distribution of model parameters and the distribution type of measurement noise. In this paper, it is assumed that the prior distribution is given as a uniform distribution with a lower- and upper-bound. It is also assumed that the measurement noise is a Gaussian distribution; that is $\varepsilon \sim N(0, s^2)$, where $s$ is the standard deviation of noise. However, users can change these assumptions easily. For example, the case when noise in data follows a lognormal distribution is considered in the crack growth problem in Section 4. In most cases, since the standard deviation of noise is unknown, it should be a part of unknown model parameters. In the case of the battery model, therefore, the vector of unknown model parameters is defined as $\Theta=\{b,s\}$. By assuming that the two model parameters are statistically independent, the prior joint PDF of the two parameters can be defined as

$$f_\Theta(\theta) = f(b)\times f(s), \quad f(b)\sim U(0, 0.05),\ f(s)\sim U\left(10^{-5}, 0.1\right) \tag{9}$$

Once the prior distribution is determined, it is necessary to build the likelihood function using the measurement data and to yield the posterior distribution shown in Eq. (8). The meaning of the likelihood function is the PDF value of obtaining the measured data $y_k$ for given model parameters $\theta=\{b,s\}$. Since the measured data are fixed, the likelihood function is a function of model parameters, which makes the likelihood function different from the PDF. If the model prediction is close to the measured data, then the likelihood is large, while the likelihood is small when the two values are significantly different. In order to build the likelihood, it is necessary to measure degradations at different times. Since the measured degradation data $(y_k, t_k)$, $k = 1, 2, ..., N_{\text{data}}$ are given at discrete times, the degradation model is also evaluated at the same discrete times as $\tilde{y}_k(b) = \tilde{y}(t_k; b)$, $k = 1, 2, ..., N_{\text{data}}$. Since the times between the measurement and the model are synchronized, $\tilde{y}_k(b)$ is only a function of model parameter $b$. The measured data $y_k$ include the random noise that is governed by $s$, while the model prediction $\tilde{y}_k(b)$ depends on $b$. Then, the likelihood function of the $k$-th measured data can be defined as

$$f_Y(y_k|\theta) = \frac{1}{s\sqrt{2\pi}} \exp\left[-\frac{1}{2s^2}\left(y_k - \tilde{y}_k(b)\right)^2\right], \quad k = 1, 2, ..., N_{\text{data}} \tag{10}$$

As shown in Eq. (8), the likelihoods of multiple data can be multiplied to obtain the posterior distribution. With $N_{\text{data}}$ data, $\mathbf{y} = \left\{y_1, y_2, ..., y_{N_{\text{data}}}\right\}$, the posterior joint PDF can be calculated by multiplying all likelihood functions and the prior PDF as

$$f_\Theta(\theta|\mathbf{y}) = \frac{1}{Ks^{N_{\text{data}}}} \exp\left[-\frac{1}{2s^2}\sum_{k=1}^{N_{\text{data}}}\left(y_k - \tilde{y}_k(b)\right)^2\right] f_\Theta(\theta) \tag{11}$$

where $K$ is again a normalizing constant.

### 2.3 Markov chain Monte Carlo sampling

Bayesian parameter estimation in Eq. (11) shows the functional expression of the posterior joint PDF of unknown model parameters. When the prior and posterior are conjugate, the posterior distribution can be expressed in the form of a standard probability distribution. In general cases, however, the posterior distribution can be expressed as a product of complex functions, such as the posterior PDF shown in Eq. (11).

The posterior PDF is then used to calculate the degradation trend and predict the RUL. For complex nonlinear models, it is difficult to propagate uncertainty in the parameters to the degradation model. Instead, samples of model parameters are generated from the posterior distribution, and the degradation model with the threshold in Eq. (2) is used to propagate these samples to calculate the samples of the end of life, and thus, the samples of the RUL in Eq. (3). Therefore, it is important to generate samples that follow the posterior distribution of parameters.

In general, the inverse cumulative distribution function (CDF) method is the easiest way of generating samples from a non-standard probability distribution, but it requires the functional expression of CDF, not PDF. For practical engineering applications, it is likely that the posterior PDF may be different from a standard probability distribution, or the posterior PDF is complicated due to the complex correlation structures between parameters. In such a case, sampling-based methods can be used to generate samples of parameters. There are many sampling methods, such as the grid approximation [15], rejection sampling [16], importance sampling [17], and the Markov Chain Monte Carlo (MCMC) method [18]. In this paper, the MCMC method using the Metropolis-Hastings (MH) algorithm is employed. MCMC is a simulation technique used to estimate quantities of interest by sampling consecutive random variables wherein the future state depends only on the current state [19].

The MCMC sampling method uses a Markov chain model in a random walk, where the distribution of the next sample depends only on the current sample (see **Figure 3**). As the algorithm generates more and more samples, the samples more closely approximate the posterior PDF. Specifically, Starting with an arbitrary initial sample (current sample), a new candidate sample is drawn from a proposal

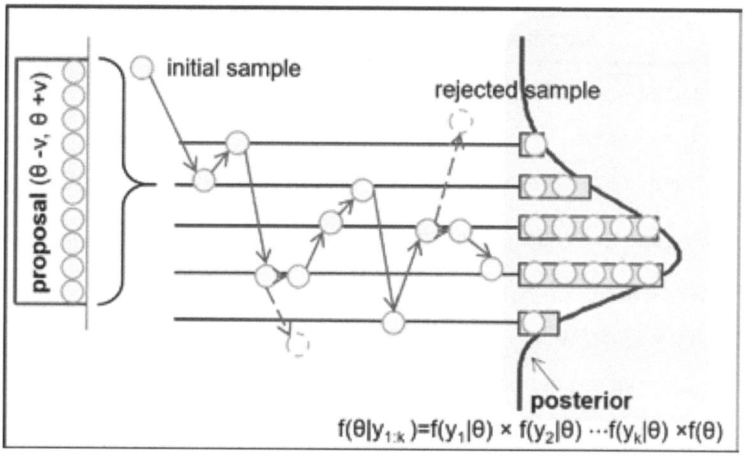


**Figure 3.**
*Markov chain Monte Carlo sampling using random walk.*

distribution centered at the current sample. In this paper, a uniformly distributed proposal distribution is used. Therefore, it is expected that the users provide the initial sample of parameters and the width of the proposal distribution. At $i$-th iteration, it is expected that the current sample $\theta^{(i-1)}$ is available, and the new candidate sample $\theta^*$ is drawn from the following proposal distribution that is uniformly distributed centered at $\theta^{(i-1)}$:

$$g\left(\theta^*, |\theta^{(i-1)}\right) \sim U\left[\theta^{(i-1)} - \mathbf{w}, \theta^{(i-1)} + \mathbf{w}\right] \quad (12)$$

where $\mathbf{w}$ is the user-provided width of the proposal distribution. It is noted that the proposal distribution is symmetric; that is $g\left(\theta^*, |\theta^{(i-1)}\right) = g\left(\theta^{(i-1)}, |\theta^*\right)$.

Once the candidate sample is generated, it is either accepted as a new sample or rejected based on an acceptance criterion. When accepted, the candidate sample is added to a new sample and used in the next iteration. When rejected, the candidate sample is discarded, and the current sample is reused in the next iteration. In the original MH algorithm, it is suggested to use a function that is proportional to the posterior distribution for the acceptance/rejection test. In this paper, however, the posterior distribution is directly used as its evaluation is not computationally expensive. Since the proposal distribution is symmetric, the following acceptance ratio can be defined:

$$Q\left(\theta^{(i-1)}, \theta^*\right) = \frac{f_\Theta(\theta^*|\mathbf{y})}{f_\Theta(\theta^{(i-1)}|\mathbf{y})} \quad (13)$$

The acceptance ratio compares the posterior probability of the new candidate sample against that of the current sample. If the candidate sample has a higher probability than that of the current sample; i.e., $Q\left(\theta^{(i-1)}, \theta^*\right) > 1$, then it is always accepted as a new sample. When $0 < Q\left(\theta^{(i-1)}, \theta^*\right) < 1$; that is, the probability of the candidate sample is lower than that of the current sample, the acceptance is determined based on the ratio. A high acceptance ratio has a more probability to be accepted, while a low ratio is occasionally accepted. This can be achieved by generating a sample from a uniform distribution, $u \sim U[0, 1]$, and the candidate sample is accepted if $u < Q\left(\theta^{(i-1)}, \theta^*\right)$; otherwise, it is rejected. Intuitively, this is why this algorithm works, and returns samples that follow the desired distribution $f_\Theta(\theta|\mathbf{y})$. In **Figure 3**, two

**Algorithm: Metropolis-Hastings algorithm**

Initial sample $\theta^{(0)}$ and posterior $f_\Theta(\theta^{(0)}|\mathbf{y})$

For i = 1 to Ns

  Sample from proposal dist. $\theta^* \sim g(\theta^*|\theta^{(i-1)})$ and posterior $f_\Theta(\theta^*|\mathbf{y})$

  Acceptance ratio: $Q(\theta^{(i-1)},\theta^*) = \dfrac{f_\Theta(\theta^* \mid \mathbf{y})}{f_\Theta(\theta^{(i-1)} \mid \mathbf{y})}$

  Acceptance probability $u \sim U(0,1)$

  If $u < Q(\theta^{(i-1)},\theta^*)$

    Accept the candidate sample: $\theta^{(i)} = \theta^*$

  otherwise

    Reject the candidate sample: $\theta^{(i)} = \theta^{(i-1)}$

  end

end

**Figure 4.**
*Metropolis-Hastings algorithm for generating samples from a posterior distribution.*

dashed circles mean that these candidate samples are not selected according to the criterion. In such a case, the current sample is selected again. This process is repeated as many times as necessary until a sufficient number of samples are obtained. **Figure 4** summarizes the MCMC sampling procedure using the MH algorithm.

The performance of MCMC sampling depends on the initial sample and the selection of the proposal distribution. A too-narrow proposal distribution can yield destabilization by not fully covering the posterior distribution, while a too-wide distribution can yield many duplications in sampling result by not accepting new samples. In addition, if the initial sample is located far away from the posterior distribution, many iterations (samples) will be required to converge to the posterior distribution. To prevent the effect of inaccurate initial samples, an initial portion of the samples can be discarded in estimating the posterior distribution, which is called the burn-in. In this paper, the first 20 percent of the samples are discarded as a burn-in.

### 2.4 Prognostics

Once the samples of parameters are obtained based on the posterior distribution, the future damage state and the RUL can be predicted by substituting the samples of parameters in the degradation model and estimating the RUL using Eqs. (2) and (3). In general, since the degradation model is a nonlinear implicit function of time, solving for $t_{\mathrm{EOL}}^{(i)}$ with a given sample $b^{(i)}$ in Eq. (2) may require an iterative process. Instead, in this paper, the degradation model is evaluated at a set of discrete future times, and then, the end of life is calculated using a simple interpolation. More specifically, let the set of discrete times is defined as

$$\text{time} = [t_0 \ t_1 \ \cdots \ t_{\mathrm{CUR}} \ \cdots \ t_{\mathrm{end}}] \tag{14}$$

Then, the measurement data are available between $t_0$ and $t_{\mathrm{CUR}}$. Bayesian parameter estimation in the previous section uses the measurement data between $t_0$ and

$t_{CUR}$ to estimate the posterior PDF of model parameters. Using the estimated model parameters, the degradations in the future times between $t_{CUR}$ and $t_{end}$ are calculated in the prediction stage. If two consecutive degradations cross the threshold; that is,

$$\left(\tilde{y}_k(b) - y_{\text{threshold}}\right) \cdot \left(\tilde{y}_{k+1}(b) - y_{\text{threshold}}\right) \le 0 \tag{15}$$

then $t_{EOL}$ exists between $t_k$ and $t_{k+1}$, which can be found by a simple linear interpolation. When the set of futures times do not include the end of life, it can extrapolate based on the trend of data. It is also possible that the degradation model never reaches the threshold; that is, the system has an infinite of life. In such a case, the sample is deleted from the calculation. Once the samples of the end of life are obtained, the samples of RUL can be calculated using Eq. (3).

Once the samples of RUL are available, the confidence interval and/or the prediction interval is used to evaluate the accuracy or precision of the RUL. The confidence interval represents how good the RUL is. Therefore, the confidence interval of 95% means that the true RUL will be within this interval with the probability of 95%. That is, the confidence interval tells us about the likely location of the true RUL. In the case of RUL samples, the 95% confidence interval can be calculated by taking the lower 2.5 percentile and the upper 2.5 percentile from the samples. On the other hand, the prediction interval shows the possible location of the next sample. Knowing that the next sample will have additional randomness from the predicted RUL, the prediction interval is calculated by adding additional randomness to the data. In practice, the RUL estimation is important for scheduling maintenance. Therefore, only the lower confidence/prediction bound is of interest in the practical application. **Figure 5** shows a representative result of prognostics, which shows the statistical distribution and the confidence interval of the RUL.

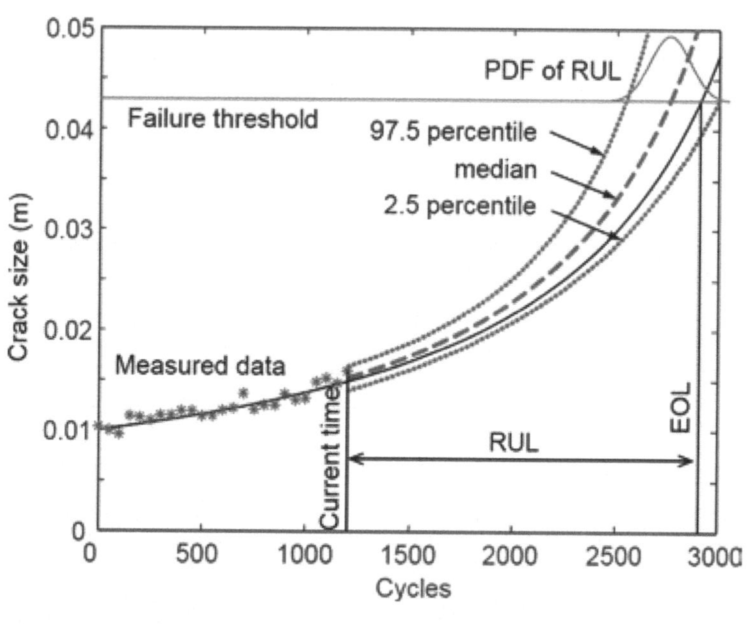


**Figure 5.**
*Statistical distribution and the confidence interval of the remaining useful life.*

## 3. MATLAB implementation

In this section, MATLAB implementation of prognostics using the Bayesian method is discussed. In the following explanation, 'line' or 'lines' in a parenthesis indicated the number of the line of the code in Appendix. The code is divided into three parts: (1) Problem Definition (lines 2–15, 65–67) (2) Bayesian parameter estimation and MCMC sampling (lines 16–29, 60–76) (3) Post-processing for displaying results (lines 40–57). Only the Problem Definition part needs to be changed for different applications. Detailed explanations are given in the subsequent sections with an example of battery degradation.

It is expected that the MATLAB script is saved as a file with the name of '**BM.m**', which has two input arguments, **para0** and **weigh** (line 1). The first argument, **para0**, is the initial sample of model parameters, and **weigh** is the width **w** of the proposal distribution in Eq. (12). The size of each array should be the same as the number of model parameters. The following is an example of calling the code in the command window of MATLAB:

```
samplResul=BM([0.011 0.02]', [0.001 0.003]');
```

In the above MATLAB command, **para0 = [0.011 0.02]'** is the initial sample of $b$ and $s$, and **weigh = [0.001 0.003]'** is the width of proposal distribution of $b$ and $s$. Since the convergence and accuracy depend on these two variables, it is suggested to try with different values. Since the users know the prior distribution of the model parameters, it is a good practice to start with the mean of the prior distribution as an initial sample. If the code ran successfully, it will return the samples of model parameters in the **samplResul** array and will generate two plots. The first plot is the trace of MCMC samples, and the second plot is the histogram of the RUL.

### 3.1 Problem definition (lines 2–15, 65–67)

The problem definition means defining the degradation equation using model parameters and time/cycle. All known parameters, as well as the initial estimate of unknown parameters, parameter names, and model data, need to be defined. The problem definition consists of two parts: parameter definition and model definition. For parameter definition (lines 2–15), '**Battery**' is used as a **WorkName** (line 3). The capacity is measured every week, so **TimeUnit** is '**weeks**' (line 4). In line 5, **time** is an array of discrete times in the units of **TimeUnit**. Measurements and predictions will be done at these times. The relative C/1 capacity data is stored as **measuData** (lines 6–7), which has 10 weeks of measurement; that is, $N_{\text{data}} = 10$ (k1 in line 18). These data correspond to the first 10 times in **time** array. Since time starts from 0, the 10th time corresponds to 9 weeks, which is the current time; that is, $t_{\text{CUR}} = 9\,\text{week}$. The degradation will be predicted for future times that are from 10 to 50 weeks. The measurement data are generated using the MATLAB script in Section 2.1. It is noted that due to random noise, the users may experience different realizations of measurement data.

In line 8, the degradation threshold $y_{\text{threshold}}$ is defined using variable '**thres**' with the value of **0.7. ParamName** in line 9 is the name of model parameters: model parameter '**b**' and standard deviation of noise '**s**'. Since the parameter name will actually be used in the model, it is important to use the actual name of variables here. The number of parameters is stored in the variable '**p**' in line 17. It is noted that the last parameter should be the standard deviation of noise '**s**'. **prioDisPar** stores the information of the prior distribution of model parameters, as given in

Eq. (9). When a uniform prior distribution is used, each row contains the lower- and upper-bounds of the distribution.

During MCMC sampling, the number of samples **Ns** is set to **5000** (line 12). Since the sampling process takes many iterations to converge, the initial 20% of the samples are discarded in calculating the posterior distribution by command **burnIn = 0.2** (line 13). In order to keep 5000 samples, **Ns/(1-burnIn)** = 6250 samples (line 21) are generated first, and then, **nBurn** = 1250 samples (line 30) are discarded.

Since the RUL is represented by **Ns** samples, the confidence interval or the prediction interval is often used to support the decision-making process. **signiLevel** (line 14) is the significance level of this interval in percentage. When **signiLevel** = 5, the code will return the lower 5 percentile, median, and the upper 5 percentile. Following is the sample output from **BM.m**:

```
# Percentiles of RUL at 9 cycles
5prct: 18.7182, median: 20.381, 95prct: 22.1576
```

For the model definition, the degradation model $\tilde{y}(t;b)$ in Eq. (1) is defined in lines 65–67. In this equation, $t$ is the time of measurement, and $b$ is the model parameter as defined in line 9. The model equation needs to be defined in such a way that component-by-component operations are possible. This is because the time and model parameters can be an array of samples.

### 3.2 Bayesian parameter estimation with MCMC (lines 16–31)

In the Bayesian parameter estimation process, the posterior distribution is expressed in terms of the product of the prior distribution and the likelihoods of all measured data. In the MATLAB code **BM.m**, the degradation model and the posterior distribution are calculated in the function **BMappl** (lines 60–76). First, the parameter samples in **param** are assigned to the variables using the **eval** function (line 62–64):

```
for j=1:size(param,1)
eval([ParamName(j,:) '=param(j,:);']);
end
```

In the case of the battery example, this command is equivalent to.

```
b = param(1,:);
s = param(2,:);
```

This is why the **ParamName** in line 9 must have the same name with the actual variable. Then, these parameters are used to calculate the degradation model with given time $t$ (line 66).

If measurement data (**measuData**) is empty, then **BMappl** only calculates the degradation model with given parameter samples at a given time $t$. This corresponds to propagating the degradation model using parameter samples to the future time for prognostics. If measurement data are provided (lines 71–74), then **BMappl** calculates the values of the posterior distribution at the parameter samples. In this case, the time should be given as an array of $N_{\text{data}}$ components from the start time to the current time. The posterior distribution in Eq. (11) (line 74) is the multiplication of the prior distribution (line 71) in Eq. (9) with the likelihood of all measured data (lines 72–73) in Eq. (10). The calculated posterior distribution from **BMappl** is used for calculating the acceptance ratio (line 23).

MCMC sampling using the MH algorithm starts with the initial sample that is provided by the users (line 19) and the value of the posterior PDF (line 20). In the loop of MCMC sampling (lines 21–29), a candidate sample is randomly generated from a uniform distribution, centered at the current sample and the width of ±**w** (line 22). The value of posterior PDF for the candidate sample is also calculated (line 23). If the acceptance ratio in Eq. (13) is greater than a randomly generated number $u$, then the candidate sample is accepted as a new sample (lines 25–26). Otherwise, the current sample and its posterior PDF are kept as a new sample and PDF. Once the MCMC sampling loop is over, the first 20% of the samples are discarded as a burn-in process (line 31). At the end of Bayesian parameter estimation, **samplResul** array contains **Ns** samples of model parameters.

### 3.3 Remaining useful life prediction (lines 32–39)

Once the samples of model parameters are obtained based on the posterior distribution, they can be used to find the RUL, which is the time when the degradation prediction reaches the threshold. First, for all samples in **samplResul**, the degradation is predicted in the future times between $t_{\mathrm{CUR}} = \mathrm{k1}$ and $t_{\mathrm{end}}$.

```
for k=1:length(time(k1:end))
  [degrPreCon(k,:),~]=BMappl(samplResul,ParamName,time(k1-1+k),[],[]);
end
```

Once the degradations in the future times are calculated, MATLAB function **interp1** is used to find the time when $\tilde{y}(t_{\mathrm{EOL}}; b) = y_{\mathrm{threshold}}$. Once $t_{\mathrm{EOL}}$ is found, the RUL can be calculated using Eq. (3).

```
for i=1:Ns
  RUL(i)=interp1(degrPreCon(:,i),time(k1:end),thres,'pchip') - time(k1);
end
```

The option 'pchip' in **interp1** uses a shape-preserving piecewise cubic interpolation, which preserves $\mathrm{C}^1$-continuity.

### 3.4 Postprocessing (lines 40–58)

In the postprocessing stage, the results given in samples are interpreted in terms of statistical quantities or in the form of graphs. First, in the **RUL** array, those components that have an infinite life should be removed (line 41). Then, the confidence intervals of [5%, median, 95%] are calculated from the **RUL** array and stored in **rulPerce** (line 42–43).

The MATLAB code plots two figures. The first figure plots the trace of MCMC samples (lines 45–50) as shown in **Figure 6(a)**. This trace shows the quality of MCMC samples. If samples are distributed randomly and symmetrically around the mean with a constant bound, then it means that the samples are stabilized and well represent the posterior distribution. If the trace of samples shows an irregular behavior as shown in **Figure 6(b)**, the samples may not represent the posterior PDF properly. This can happen when the initial sample was far away from the mean and when the width of the proposal distribution is too narrow or too wide.

The second plot is the histogram of RUL (lines 51–53) as shown in **Figure 7**. At the end of the code, the confidence intervals of [5%, median, 95%] are printed on the command window (lines 54–56). All variables are saved in the computer file so that they can be loaded to the memory for further analysis (line 57). The name of the saved database is "WorkName at $t_{\mathrm{CUR}}$.mat". For example, in the battery case, the saved database name is "Battery at 9.mat".

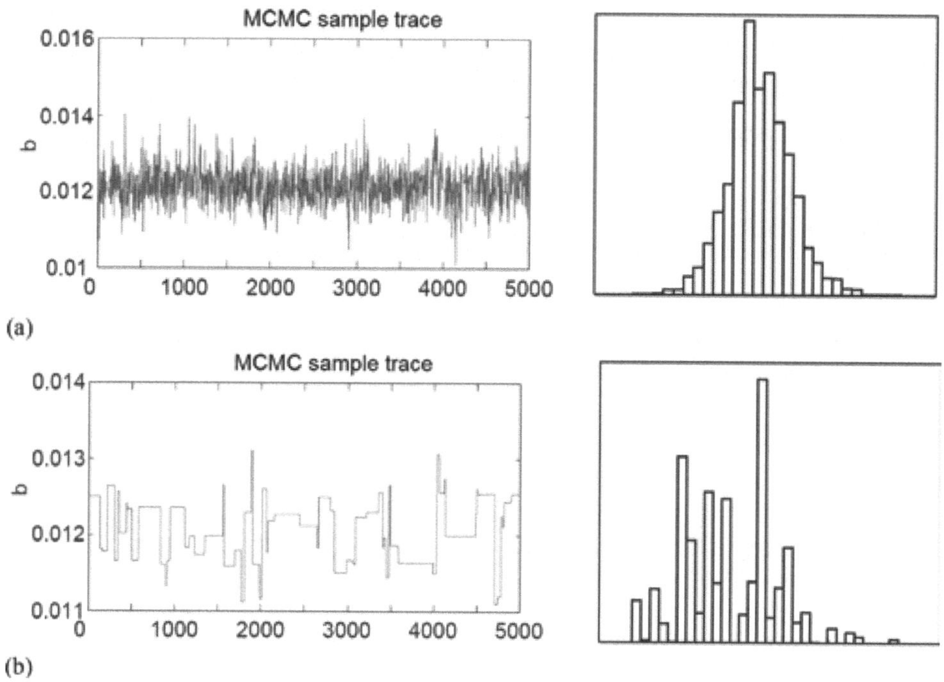


**Figure 6.**
*Trace of samples from MCMC sampling (a) proper samples and (b) improper samples.*

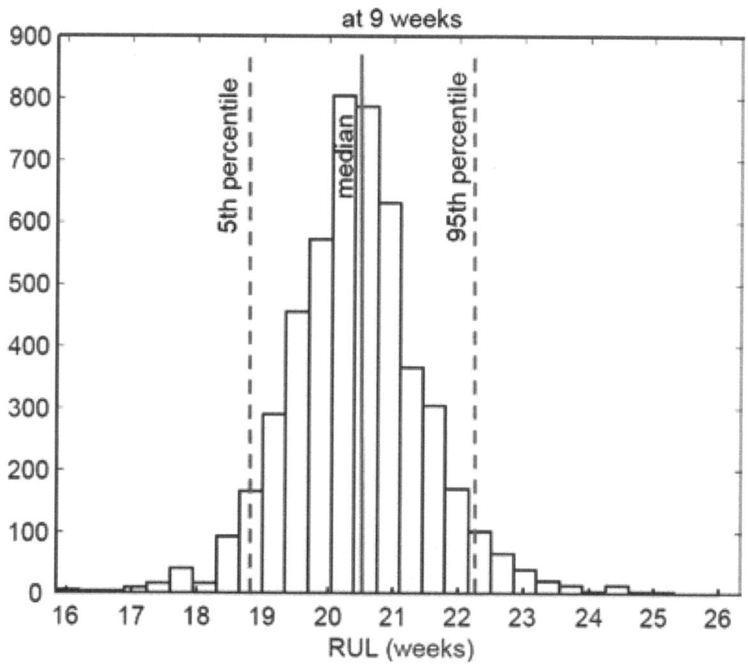


**Figure 7.**
*Histogram of the remaining useful life.*

Although the MATLAB code plots two figures, it is possible that the users can plot different figures using the saved database. After calling the **BM.m** function, the saved database has to be loaded to the memory using the following commands:

```
clear; clc; load('Battery at 9.mat')
```

In addition to the trace of samples shown in **Figure 6**, it is possible to plot the histogram of model parameters using the same samples. The following commands plot the histograms of all model parameters.

```
figure; % histogram of parameters
for j=1:p;
  subplot(1,p,j);
  hist(samplResul(j,:),30);
end
```

When the true degradation model is known, it is possible to compare the predicted degradation with the true one. The following MATLAB script plots the median and confidence intervals of the predicted degradation along with the true degradation and the threshold.

```
figure;
degraTrue=exp(-0.012.*time);
degraPI=prctile(degraPredi',perceValue)';
plot(time,degraTrue,'k'); hold on;
plot(time(1:k1),measuData,'*b');
plot(time(k1:end),degraPI(:,1),'-r');
plot(time(k1:end),degraPI(:,2:3),':r');
plot([0 time(end)],[thres thres],'g');
xlabel('weeks'); ylabel('Relative C/1 capacity');
```

The degradation curves up to 50 weeks are shown in **Figure 8** for the battery example. The true degradation with $b = 0.012$ is shown with the black curve. The red curves show 5, 50 (median) and 95 percentiles of the predicted degradation, which are caused by **signiLevel** = 5 (line 14). The plot also shows the threshold (green line) and measurement data (blue asterisk marks). Based on the true model, the end of life of the battery is $t_{\mathrm{EOL}} = 29.72$ weeks, and thus, the true RUL should be $t_{\mathrm{RUL}} = 20.72$ weeks. The prediction shows that the median of RUL is

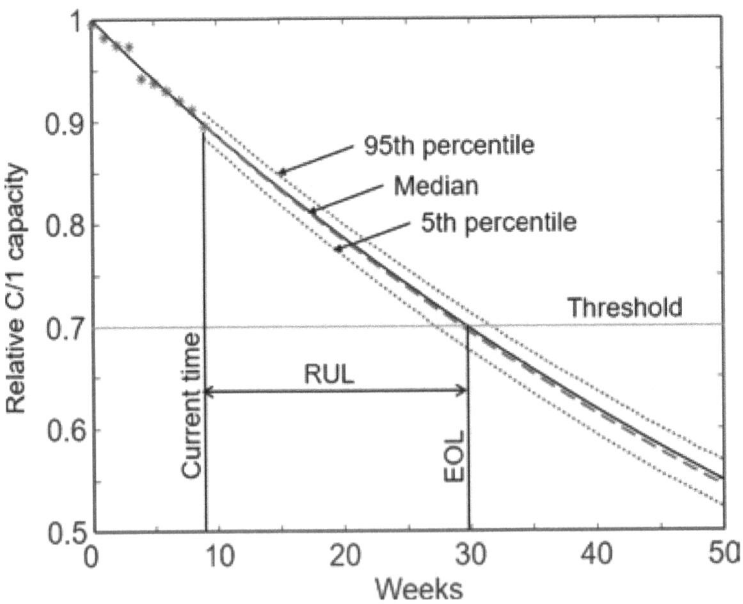


**Figure 8.**
*Comparison of the predicted degradation with the true degradation.*

20.38 weeks, which is close to the true RUL. In addition, the 90% confidence interval is about 3.4 weeks; that is, the uncertainty in the prediction is about 17%.

## 4. Application to crack growth prognostics

The code can be modified easily by the users for various applications. In this section, an example of crack growth is used to explain how the MATLAB code **BM.m** can be modified.

### 4.1 Model definition: crack growth

In fatigue crack growth, the failure criterion is given in terms of the crack size. Therefore, it would be appropriate to use the size of crack as a degradation feature. In this case, the degradation feature monotonically increases, while it was monotonically decreased for the battery example. Assuming that a through-the-thickness center crack exists in an infinite plate under mode-I loading condition, the rate of fatigue crack growth can be expressed using the Paris-Erdogan model as

$$\frac{da}{dN} = C(\Delta K)^m \tag{16}$$

where $a$ is the half crack size, $N$ is the number of cycles, $m$ and $C$ are model parameters, $\Delta K = \Delta\sigma\sqrt{\pi a}$ is the range of stress intensity factor, and $\Delta\sigma$ is the stress range. It is assumed that time is the number of fatigue loading cycles. For the consistent notation, the crack size and cycles are replaced with $\tilde{y} = a$ and $t = N$ in the following explanation. Since the degradation model requires the crack size as a function of time and model parameters, Eq. (16) can be integrated to obtain the following degradation model:

$$\tilde{y}(t; m, C) = \left[tC\left(1 - \frac{m}{2}\right)\left(\Delta\sigma\sqrt{\pi}\right)^m + a_0^{1-\frac{m}{2}}\right]^{\frac{2}{2-m}} \tag{17}$$

The system is under fatigue loading with the range of stress being $\Delta\sigma = 75\text{MPa}$ at each cycle. It is assumed that the health monitoring is performed every 50 cycles to measure the crack size $y_k$ until the current time $t_{\text{CUR}} = 1,200$ cycles, and the initial size of the crack is $a_0 = 0.01\text{m}$. Similar to the battery degradation example, the measurement data are simulated by adding random noise to the true crack size. First, the true crack size data are generated at every 50 cycles using Eq. (17) with $m_{\text{true}} = 3.8$ and $C_{\text{true}} = 1.5 \times 10^{-10}$. The measured crack size data are then generated by adding Gaussian noise $\varepsilon \sim N(0, s^2)$, $s = 0.0005\text{m}$ to the true crack sizes. The measured crack size data are used to identify three model parameters, $\theta = \{m, \ln(C), \ s\}$. In the Paris-Erdogan model, the y-intercept $C$ is very small but changes its magnitude by several orders. Therefore, it would be better to identify logarithm of $C$. For RUL calculation, the critical crack size is determined as 0.043 m.

For the Bayesian method, it is necessary to define the prior distribution and the likelihood function. In the battery example, it was assumed that noise in data follows a normal distribution. However, when the distribution type of measurement noise is unknown, it is possible that the likelihood function might be different from the true noise distribution. The same is true for the prior/initial distribution. Therefore, it would be a good exercise to study the effect of different distribution types by changing the MATLAB codes. In this example, the lognormal distribution is employed for the likelihood function as:

$$f(y_k|\theta) = \frac{1}{y_k \zeta_k \sqrt{2\pi}} \exp\left[-\frac{1}{2}\left(\frac{\ln y_k - \eta_k}{\zeta_k}\right)^2\right], \quad k = 1, \dots, N_{\text{data}} \tag{18}$$

where $\zeta_k = \sqrt{\ln\left[1 + \left(s/\tilde{y}_k\right)^2\right]}$ and $\eta_k = \ln\left(\tilde{y}_k\right) - 0.5(\zeta_k)^2$ are the standard deviation and mean of a lognormal distribution, respectively. In the above equation, $\tilde{y}_k(\theta)$ is the model prediction from Eq. (17) at time $t_k$ with given model parameters $m$ and $C$.

Also, the prior distribution of each parameter is assumed as a normal distribution as $f(m) \sim N(4, 0.2^2), f(\ln C) \sim N(-22.33, 0.5^2)$, and $f(s) \sim N\left(5 \times 10^{-4}, \left(1 \times 10^{-4}\right)^2\right)$. Therefore, the joint prior distribution can be obtained from the independence assumption as

$$f_\Theta(\theta) = f(m) \times f(\ln C) \times f(s) \tag{19}$$

The posterior distribution can be obtained by multiplying the prior distribution in Eq. (19) with the likelihood function in Eq. (18).

### 4.2 Modifying the code

For the crack growth example, the code in Appendix needs to be changed as follows. First, the problem definition part in lines 2–15 is replaced with the following code:

```
%===== PROBLEM DEFINITION 1 (Required Parameters)==================
WorkName='Crack';                          %work results are saved by WorkName
TimeUnit='cycles';                                        % time unit name
time=(0:50:3600)'; % time including both at measurement and at prediction
measuData=[1.03 1.00 0.96 1.14 1.13 1.10 1.15 1.15 1.19 1.19 1.14 ...
              1.14 1.20 1.22 1.37 1.21 1.25 1.25 1.36 1.30 1.32 1.48 ...
              1.52 1.47 1.59]'*0.01;
thres=0.043;                                  % threshold - critical value
ParamName=['m'; 'C'; 's'];          % model parameters' name to be estimated
prioDisPar=[4 0.2; -22.33 .5; 5E-4 1E-4];   % parameter prior distributions
Ns=10000;                              % num. of samples for MCMC simulation
burnIn=0.2;                                             % burn-in fraction
signiLevel=2.5;                       % significance level for C.I. and P.I.
%==================================================================
```

A total of $N_{\text{data}} = 25$ measurement data are provided up to $t_{\text{CUR}} = 1,200$ cycles, and the degradation is predicted until $t_{\text{end}} = 3,600$ cycles. Since the prior distributions are assumed a normal distribution, the first column of **prioDisPar** is the mean, and the second column is the standard deviation.

Next, the model definition part in lines 65–67 is replaced with the following code:

```
%===== PROBLEM DEFINITION 2 (model equation)=======================
a0=0.01; dsig=75; coef=1-m/2;
degraModel=(t.*exp(C).*coef.*(dsig*sqrt(pi)).^m + a0.^coef).^(1./coef);
loca=imag(degraModel)~=0; degraModel(loca)=real(degraModel(loca));
%==================================================================
```

Initially, a fatigue crack grows slowly and then grows rapidly just before becoming unstable. The crack growth model in Eq. (17) is only valid when the crack growth is stable. Normally the threshold $y_{\text{threshold}}$ is set before the crack becomes unstable. When the crack becomes unstable, Eq. (17) yield a complex number. Therefore, the last line of the model definition code identifies if a prediction yields a complex number and converts it to a real number by ignoring the complex part.

In addition to the problem definition part, the posterior distribution part also needs to be modified because instead of a uniform distribution, a normal distribution is used for the prior distribution. Also, lognormal distributions are used instead of normal distributions for the likelihood function. The posterior distribution part in lines 71–74 needs to be modified as follows:

```
prior=prod(normpdf(param,prioDisPar(:,1),prioDisPar(:,2)));
likel=1;
for k=1:length(measuData)
  zeta=sqrt(log(1+(s./degraModel(k)).^2)); eta=log(degraModel(k))-0.5*zeta.^2;
  likel=lognpdf(measuData(k),eta,zeta).*likel;
end
poste=likel.*prior;
```

### 4.3 Results

Similar to the battery example, the MATLAB code can be used to plot the trace of MCMC sampling and the histogram of model parameters and RUL. Using a similar code provided in Section 3.4, the degradation trend could also be plotted. For example, **Figure 5** shows the predicted degradation with the true degradation. Even if the median (1553 cycles) has a relatively large error with the true RUL (1709 cycles), the 95% confidence interval covers the true RUL.

An important information that was not discussed before is the correlation between model parameters. It is well known that the Paris-Erdogan model parameters, $m$ and $C$, are strongly correlated [20]. Therefore, it would be beneficial to plot the MCMC samples in the parameter space. The following MATLAB script plots the MCMC samples of the parameters.

```
clear; clc; load('Crack at 1200.mat');
[m, C]=meshgrid((3.4:0.009:4.3)',(-24:0.025:-21.5));
for i=1:101; for j=1:101
   para=[m(i,j); C(i,j); 0.0005];
   [~,post(i,j)]=BMappl(para,ParamName,time(1:k1),measuData,prioDisPar);
end; end
plot(samplResul(1,:),samplResul(2,:),'.','color',[0.5 0.5 0.5]); hold on
contour(m,C,real(post));
plot(3.8,log(1.5E-10),'pk','markersize',14);
```

**Figure 9** shows the MCMC samples in the parameter space along with the exact value of the parameters (star marker). The figure also shows the contour of the joint posterior PDF based on the grid method. It can be observed that the MCMC samples represent the joint posterior PDF well, and the joint PDF covers the true parameter values. However, due to a strong correlation between the two Paris-Erdogan model parameters, the joint PDF shows a narrow but long tail. Any combination of model parameters along the correlation line may yield a similar damage growth trend.

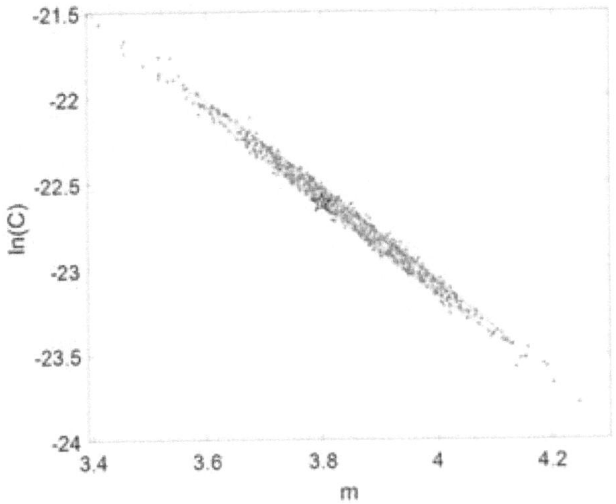


**Figure 9.**
*Correlation between the two Paris-Erdogan model parameters.*

## 5. Conclusions

This paper presents a Bayesian-based prognostics algorithm with a MATLAB code. This code is constructed with simply 76 lines in the case of a battery degradation example. Users can easily modify this code as per their own application. As an example of code modification, the case of crack growth model is also presented. The paper also provided several MATLAB scripts to help plot the degradation curve and correlation between multiple parameters.

## A. Appendix

```
function samplResul=BM(para0,weigh)
%===== PROBLEM DEFINITION 1 (Required Parameters)================
WorkName='Battery';                    % work results are saved by WorkName
TimeUnit='weeks';                                        % time unit name
time=(0:50)';   % time including both at measurement and at prediction
measuData=[0.9951 0.9826 0.9750 0.9736 0.9424 0.9381 ...
           0.9300 0.9203 0.9114 0.8952]';   % measured data at time (0:9)
thres=0.7;                                   % threshold - critical value
ParamName=['b'; 's'];            % model parameters' name to be estimated
prioDisPar=[0 0.05;                        % parameter prior distributions
            1e-5 0.1];
Ns=5000;                             % num. of samples for MCMC simulation
burnIn=0.2;                                           % burn-in fraction
signiLevel=5;                       % significance level for C.I. and P.I.
%===============================================================
%%% Bayesian parameter estimation with MCMC
p=size(ParamName,1);                                 % num. of parameters
k1=length(measuData);                                      % num. of data
sampl(:,1)=para0;                        % Initial samples of parameters
[~,jPdf0]=BMappl(para0,ParamName,time(1:k1),measuData,prioDisPar);
for i=2:Ns/(1-burnIn)                                     %% MCMC Process
```

```
para1(:,1)=para0+weigh.*(2*rand(p,1)-1); % sample from proposal dist.
[~,jPdf1]=BMappl(para1,ParamName,time(1:k1),measuData,prioDisPar);
if rand<(jPdf1/jPdf0) && jPdf1>0                    % acceptance criterion
para0=para1;
jPdf0=jPdf1;
 end
 sampl(:,i)=para0;                                        % new MCMC sample
end
nBurn=Ns/(1-burnIn)-Ns;                      % No. of effective MCMC samples
samplResul=sampl(:,nBurn+1:end);                   % Final Sampling results
%%% RUL prediction
for k=1:length(time(k1:end))                        % degradation prediction
 [degrPreCon(k,:),~]=BMappl(samplResul,ParamName,time(k1-1+k),[],[]);
 degraPredi(k,:)=degrPreCon(k,:)+normrnd(0,samplResul(end,:));
end
for i=1:Ns                                                   % RUL prediction
 RUL(i)=interp1(degrPreCon(:,i),time(k1:end),thres,'pchip') - time(k1);
end
%%% POST-PROCESSING
Index=isnan(RUL); RUL(Index)=[];
perceValue=[50 signiLevel 100-signiLevel];         % median & confidence intervals
rulPerce=prctile(RUL,perceValue);                         %percentiles of RUL
figure(1);
for j=1:p % plotting MCMC sample trace
 subplot(p,1,j); % for all model parameters
 plot(samplResul(j,:));
 ylabel(ParamName(j,:));
 title('MCMC sample trace');
end
figure(2); set(gca,'fontsize',14); hist(RUL,30);                % RUL histogram
xlim([min(RUL) max(RUL)]); xlabel(['RUL' ' (' TimeUnit ')']);
titleName=['at ' num2str(time(k1)) ' ' TimeUnit]; title(titleName)
fprintf( '\n # Percentiles of RUL at %g cycles \n', time(k1))
fprintf('\n %gprct: %g, median: %g, %gprct: %g \n' , perceValue(2), ...
           rulPerce(2), rulPerce(1), perceValue(3), rulPerce(3))
Name=[WorkName ' at ' num2str(time(k1)) '.mat']; save(Name); % save work
end
%
function [degraModel, poste]=BMappl(param,ParamName,t,measuData,prioDisPar)
% Evaluate the degradation model or posterior PDF
for j=1:size(param,1)
eval([ParamName(j,:) '=param(j,:);']);
end
%===== PROBLEM DEFINITION 2 (model equation)===================
degraModel=exp(-b.*t);
%===========================================================
if isempty(measuData)
poste=0;
else
prior=prod(unifpdf(param,prioDisPar(:,1),prioDisPar(:,2)));        % prior
likel=(1./s).^length(measuData) ...                             % likelihood
```

```
           .*exp(-0.5./s.^2.*norm(measuData-degraModel)^2);
poste=likel.*prior; % posterior
end
end
```

## References

[1] Giurgiutiu V. Structural Health Monitoring with Piezoelectric Wafer Active Sensors. 2nd Edition. Waltham, MA, USA: Academic Press; 2014

[2] Sohn H, Farrar CR, Hemez FM, Czarnecki JJ, Shunk DD, Stinemates DW, et al. "A Review of Structural Health Monitoring Literature: 1996–2001," Report Number LA-13976-MS. Los Alamos, NM: Los Alamos National Laboratory; 2004

[3] Kim NH, An D, Choi J-H. Prognostics and Health Management of Engineering Systems: An introduction. Switzerland: Springer International Publishing; 2017. DOI: 10.1007/978-3-319-44742-1

[4] Si XS, Wang W, Hu CH, Zhou DH. Remaining useful life estimation—A review on the statistical data driven approaches. European Journal of Operational Research. 2011;**213**:1-14

[5] Lee J, Wu F, Zhao W, Ghaffari M, Liao L, Siegel D. Prognostics and health management design for rotary machinery systems—reviews, methodology and applications. Mechanical Systems and Signal Processing. 2014;**42**(1–2):314-334

[6] Saha B, Goebel K, Christophersen J. Comparison of prognostic algorithms for estimating remaining useful life of batteries. Transactions of the Institute of Measurement and Control. 2009;**31** (3–4):293-308

[7] Xing Y, Williard N, Tsui K-L, Pecht M. A comparative review of prognostics-based reliability methods for Lithium batteries. In: Prognostics and System Health Management Conference, Shenzhen, China; 24-25 May 2011

[8] Zhang J, Lee J. A review on prognostics and health monitoring of Li-ion battery. Journal of Power Sources. 2011;**196**:6007-6014

[9] An D, Choi J-H, Kim NH. Prognostics 101: A tutorial for particle filter-based prognostics algorithm using Matlab. Reliability Engineering and System Safety. 2013;**115**:161-169. DOI: 10.1016/j.ress.2013.02.019

[10] Goebel KB, Saha A, Saxena JR, et al. Prognostics in battery health management. IEEE Instrumentation and Measurement Magazine. 2008;**11**(4): 33-40

[11] Guan X, Jha R, Liu Y. Model selection, updating and averaging for probabilistic fatigue damage prognosis. Structural Safety. 2011;**33**(3):242-249

[12] Bayes T, Price R. An essay towards solving a problem in the doctrine of chances. By the late rev. Mr. Bayes, communicated by Mr. Price, in a letter to John Canton, A. M. F. R. S. Philosophical Transactions of the Royal Society of London. 1763;**53**:370-418. DOI: 10.1098/rstl.1763.0053

[13] An D, Choi J-H, Kim NH, Pattabhiraman S. Fatigue life prediction based on Bayesian approach to incorporate field data into probability model. Structural Engineering and Mechanics. 2011;**37**(4):427-442

[14] Athanasios P, editor. Probability, Random Variables, and Stochastic Processes. New York: McGraw-Hill; 1984

[15] Gelman A, Carlin JB, Stern HS, et al., editors. Bayesian Data Analysis. New York: Chapman & Hall; 2004

[16] Casella G, Robert CP, Wells MT. Generalized Accept-Reject Sampling Schemes. Lecture Notes-Monograph Series. Vol. 45. Beachwood: Institute of Mathematical Statistics; 2004. pp. 342-347

[17] Glynn PW, Iglehart DL. Importance sampling for stochastic simulations.

Management Science. 1989;**35**(11):
1367-1392

[18] Andrieu C, Freitas DN, Doucet A, et al. An introduction to MCMC for machine learning. Machine Learning. 2003;**50**(1):5-43

[19] An D, Kim NH, Choi J-H. Practical options for selecting data-driven or physics-based prognostics algorithms with reviews. Reliability Engineering & System Safety. 2015;**133**:223-236. DOI: 10.1016/j.ress.2014.09.014

[20] An D, Choi J-H, Kim NH. Identification of correlated damage parameters under noise and bias using Bayesian inference. Structural Health Monitoring. 2012;**11**(3):293-303. DOI: 10.1177/1475921711424520

Chapter 3

# Predicting Sets of Automata: Architecture, Evolution, Examples of Prognosis, and Applications


## Abstract

This chapter describes the sets of interacting automata constructed on the cascades of wavelet coefficients of input signal. The basic principles of the evolution of automata during the processing of incoming cascades and the vector of processes consisting of segments of cascades of constant length are described. The main principles of constructing the family of automata are determined from the internal symmetry of incoming cascades and the definition of symmetry groups of vector processes and their isotropy groups. The trajectories of states are defined on nontrivial topological spaces, the so-called degeneration spaces of the characteristic functional. The family of evolving automata with tunable communications architecture is designed to predict the state of engineering objects and identify predictors, early predictors, and hidden predictors of failure. This chapter provides examples of the work of predictive automata in various fields of engineering and medicine. It demonstrates the operation of the automaton in spaces with a nontrivial topology of input cascades, algorithms of the predictor search, and estimations. The family of evolving automata with reconstructing architecture of connections is designed to predict the state of engineering objects and medicine and identify predictors, early predictors, and hidden predictors of failure. The architecture and functional properties of automata are determined from the results and main conclusions.



**Keywords:** preventive monitoring, failure prognosis, remote calculating cluster, optimize drug therapy, Turing machine, maintenance optimization, preventive maintenance, remaining useful life


## 1. Introduction

This chapter is devoted to a detailed analysis of the structure and properties of predictive automata sets and the analysis of their ability to predict and to achieve accurate time estimates of the predicted events.

The creation of set of predictive automata was preceded by the construction of various models for prognosis of the state of technical devices; basically it was about rotational and reciprocating mechanisms [1]. The same interests of the authors attracted research in the field of prognosis in cardiology [2, 3], where, especially in the last decade, significant progress has been made in understanding the mechanisms of the emergence of various kinds of arrhythmias and the mechanisms of so-called sudden cardiac death. In the field of technical devices, some progress is observed in the modeling of turbulent combustion of fuels.

As is known, the diagnosis and prognosis of combustion modes, their stability serves as the basis for the development of a causal prognosis, including for the mechanics of various types of engines and turbines, as well as for the analysis of hydrogasdynamic processes that support the combustion modes of fuel. Uniting all these systems is the fact that both the processes of propagation of action potential in the myocardium and the processes of combustion of fuels belong to the so-called reaction-diffusion systems. Regarding cardiological applications, it is also necessary to note the class of tasks for prognosis and management associated with implantable devices represented by the CRT and ICD devices. Here arises a problem of prognosis and developing a strategy for the management of devices in order to prevent such heart events as multiple births of sources of secondary waves in the myocardium, leading to fibrillation of the ventricles of the heart and to sudden cardiac death [4]. This chapter contains examples and comments on the operation of automata in various machines, without avoiding examples from cardiology. Demonstrating systems of automata sets for predicting objects of various natures (mechanisms and objects of biological nature), the authors sought to build automata on general principles that are universal for a wide range of objects. As an introduction to the subject matter, it is necessary to describe the basic conceptual constructions that precede the construction of set of recognizing automata and the description of their properties. We are talking about the substantive part of the hierarchy model [5] in the prognostic tasks. It is implied that the observed signals from various sensors and devices are represented as a continuous or discrete series of their wavelet coefficients [6]. All the observed signals from accelerometers, pressure sensors, and the sensors of the angle of rotation of the shaft, along with them the signals of the electrodes from the human body, have a quasi-periodic nature, i.e., their periods differ by a certain random value, for which when collecting statistics, one can use sequences of a series of signals with close periods. To simplify the presentation in the future, if there are no special reservations, only one class of wavelet coefficients generated by changing the number of the cycle or period will be considered while fixing all other indices of wavelet coefficients. Thus, each indexed cascade with numbers of cycles taken as discrete time corresponds to a cascade of wavelet coefficients with fixed indices of wavelet coefficients [6].

Further consideration of segments of a cascade of fixed length generates a cascade of vectors, where the growth of the cycle number is selected as the direction. The following construction of the task of prognosis is reduced to the representation of the task of wandering a vector along a multidimensional lattice or along its continual analog of a space of dimension $N^*$, where $N^*$ is the length of the selected segment. Further representation of the probability of transition from the initial vector to the final vector in the form of the Feynman integral along trajectories in $L$ steps leads in the continuum limit to the well-known evolution equations of the Fokker-Planck type for the probability density $P(\boldsymbol{R}_0, \boldsymbol{R}_L, L)$ [2, 7]. If the final vector is predetermined, then the problem of determining the number of steps that is necessary to achieve a given vector arises. The solution of such task gives the time to achieve a given vector; in other words, if the vector is specified on the border of the failure, then the remaining useful life is determined. If the solution of the evolution equation is known, for example, its solution in the moments, then further analytical calculation of the RUL is simple.

However, the presented approach to prognosis harbors many underwater reefs. The main difficulties are associated with the variability of properties of the observed signal and, therefore, the properties of the cascades. In terms of wandering around the lattice or in its continuum analogs, these properties are associated with the presence of certain prohibitions, for example, wandering without self-intersections or taking into account the excluded volume [8] and more realistic taking into account the prohibitions on a part of admissible trajectories connected with the physical boundary of the failure. The listed and not only prohibitions generate a state space of $R^{N^*}$ with forbidden states, which makes the state space a complex, multiply connected subset in $R^{N^*}$. In this case, the process under consideration may lose the properties of Markov process; the Chapman-Kolmogorov identity is not satisfied. Ultimately, the evolution equations become integro-differential, and moreover, there is a need to introduce many-particle densities of probability transition [7]. As a result the solution of evolution equations becomes difficult and hardly solvable for the general case.

In addition, most often, in practice, the probabilities of transition from nodes of a multidimensional lattice (elementary transitions) are unknown. The listed difficulties induce to the further development of the model free from the noted problems. For this, together with the main state space of the cascade vector, a symbolic space is constructed. Symbolic space appeals to the frequency representation of the process in the form of a histogram of a cascade vector built on each vector. In this case, elementary transitions are represented by an abelian subgroup of matrices. At the same time, multiplying such a matrix by a column from the frequencies in the representation of the affine space is reduced to adding or subtracting one in two fixed coordinates of the frequency vector. This fact allows us to consider the frequency histograms of the vectors of the cascade vector as the internal states of the Turing machine. Elementary transitions change the internal state of the Turing machine and correspond to the shift of the incoming tape by one. Some analogy with topological Markov chains is also possible, but the transition matrix contains both 1 and $-1$. The representation of the vector column of the frequency histogram in Euclidean space of dimension equal to the number of components or the length of the column allows to represent the process of changing the internal states of the automaton as a walking in $\boldsymbol{R}^N$ space, or taking into account the obvious restriction on the value of the sum of the column components to represent the same process as a walking on a multidimensional simplex $\Sigma^{n-1}$ of the dimension n$-$1 [2].

The vector column in the process of evolution of the system is described by solving the basic kinetic or balance equations [2].

The constructed sets of automata allow to predict or determine one's state at future times based on the set of states at previous moments, i.e., on the basis of knowledge of the state of the automaton and transition probabilities. Automata themselves receive this knowledge on the basis of a set of statistics after entering the stationary mode. The next step in the formalization and algorithmization of the prognostic tasks is to describe the set of admissible values of the state vectors. To this end, by analogy with the theory of topological defects of condensed media, a degeneration space $G/H$ and k-multiple loop space $\Omega^k(G/H)$ of degeneration space are introduced. Due to this step, the formalization of the prognosis model is completed, and further in the work, a lot of the automata for the prognosis are constructed.

## 2. Description of topological dynamics

The immediate goal of this section is to describe an automaton operating in a homogeneous space or space of degeneration. This section completes the construction of the set of interacting automata. To construct the set of automata defined for each cascade of wavelet coefficients of the observed signal, a certain symbolic space was defined, which allows treating the change in the internal states of the automaton as a wandering point on an n-1-dimensional simplex. In parallel with this, the evolution of the states of the automaton was interpreted as the wandering of a point in the multidimensional homogeneous space or the degeneration space $G/H$. The construction given in [6] without taking into account the topology of the space of degeneracy leads to evolution equations such as the balance equation or the master equation and, in more general cases, to the basic kinetic equation. The only restriction imposed on the probabilities of transitions from the quasi-stationarity condition was reduced to the fulfillment of the quasi-stationarity conditions, imposed on transition probabilities between histogram columns [2]:

$$\sum_k \hat{w}_{k,m} = \sum_m \hat{w}_{m,k} \tag{1}$$

The transition from the initial state to the final state was determined by the product of elementary transitions, represented as matrices acting in affine space, and was reduced to subtracting one from one component of the vector to adding one to another component. Some analogy with topological Markov chains was noted.

This chapter is a direct continuation of the presentation of the models and algorithms given in the works [2, 3]. The basic model is reduced to the construction of two main spaces. The first of them is the state space, and it is defined as a subset in $R^{N^*}$ of all admissible values of the wavelet segments of the coefficients of the observed signal at the fixing all indices of wavelet coefficients ${}_{Hist}^{\;k}W_{i,j}^{N}$, except the quasi-period index $N$.

Necessary definitions:

Cascade definition:

$$\left\{ {}_{Hist}^{\;k}W_{i,j}^{N}; N = 1, 2, 3, .....\infty \right\} \tag{2}$$

Defining segments and defining states:

$$\left\{ {}_{Hist}^{\;k}W_{i,j}^{N}; N = 1, 2, 3, ..., N^* \right\} \in R^{N^*} \tag{3}$$

Definition of trajectories:

$$\{\boldsymbol{R_k}\} = \left\{ {}_{Hist}^{\;k}W_{i,j}^{N} : k\, N^* \leq N \leq (k+1)N^*, k = 0, 1, 2, 3... \right\}, \forall \boldsymbol{k}, \boldsymbol{R_k} \in \boldsymbol{R}^{N^*} \tag{4}$$

On the cascades thus defined, segments of fixed length are defined. The segment length can be different, but the main requirement for the length is that at this length, the cascade goes into a quasi-stationary mode. That is, the following assumption is implicitly admitted: it follows from the construction that each cascade is indexed by the numbers of the columns of the histograms of quasi-periods, that is, each cascade corresponds to the numbers of the periods of almost constant length. For example, for rotary equipment, the set of revolutions of the shaft is indexed by the sequence numbers of revolutions with some fixed time of complete rotation. Thus, the set of revolutions is factorized by a histogram of the shaft revolution duration. For four-stroke internal combustion engines, the shaft turning time is fixed at $4\pi$. In cardiac applications, a histogram of beat-to-beat durations, etc. is determined. In part, this approach solves the synchronization problem by allowing the signal to be represented on a quasi-period as the implementation of some random process.

In this case, the assumption of the exit mode to quasi-stationarity in some cases, for example, if the process is diffusive, presupposes the existence of such a regime not only during the exit but also with the further evolution of the stochastic process. For example, in the number of additional conditions, the Lyapunov functional is determined and an analog of the $H$-theorem of Boltzmann is valid. In this case, the Lyapunov functional is entropy, which is a complete analog of the Kulbak entropy. At the same time, the nonstationary density of the distribution function asymptotically tends monotonically to the stationary one:

$$\boldsymbol{H} = \int_a^b dx p(x,t) \ln \left[ p(x,t)/p_s(x) \right] \tag{5}$$

The next step comes down to a more detailed description of the subset of states, i.e., sets of admissible states in $\boldsymbol{R}^N$.

The basic principle of the construction of the marked set appeals to the construction of the degeneration spaces of the free energy functional in various condensed media [9]. For this, the symmetry group $G$ for the generating functional is determined, and then the isotropy group of the state vector is determined. The factorization of a symmetry group $G$ into isotropy subgroups $H$ gives the so-called degeneration space $G/H$. In essence, the symmetry group transitively acts where the states of the automaton defined in [2] are invariant with respect to the action of group $G$; the transition probability is constant during transformations of the state vector by multiplying the elements by the action of a symmetry group.

For further formalization of the prognosis task, all homotopy classes of the degeneration space $G/H$: $[M, G/H]$ are considered. Mapping $M \to G/H$ defines the set of system states. In particular, as $M$ spheres of various dimensions $S^i$ are considered. In this case

$$\left[ S^i, G/H \right] = \pi_i(G/H) \tag{6}$$

And $i = 1$, $\pi_i(G/H)$—a group, $i > 1$, $\pi_i(G/H)$—an Abelian group.

Some examples of the construction of degeneration spaces are described in this paper [10]. It is shown that under the assumptions made, the degeneration space for some set of segments of a fixed length is the so-called homogeneous space. The term "degeneration space" is often used in physics, which is more accurate, since with respect to the action of degeneration groups on vector states, they are invariant with respect to the characteristic functional of the process. In other words, it is assumed that the states of the predictive automaton introduced in the cited paper are invariant with respect to the discrete subgroup of the translation group. In this case, translation is carried out along a cascade with a step $N^*$. There is some analogy to the constructed objects with stochastic processes with a measure invariant with respect to the group of translations [11].

Since the properties of the signal are initially unknown, and under the assumption that the properties of the signal may change, it turns out that it is necessary to consider nested sequences by dimension:

$$R^{N_1} \subset R^{N_2} \subset R^{N_3} \subset \ldots \subset R^{N_k} \subset; \quad G_1/H_1 \subset G_2/H_2 \subset \ldots \subset G_k/H_k \subset \ldots \tag{7}$$

Nested sequences of groups and spaces of degeneracy, these sequences are constructed for each index of the wavelet coefficients; therefore, all trajectories generated by such embeddings as multitrajectories are considered.

The analog of the Turin machine defined in the work [2], hereinafter referred to as the recognizing automaton, moving along the cascade, changes its internal state. A change in the internal state generates a trajectory in the $n_k - 1$-dimensional simplex $\Sigma^{n_k-1}$. Accounting for all coefficients of the wavelet decomposition of the signal thus generates a set of predictive automata. Further, it is shown that the change in the internal state of the automaton is reduced to elementary steps on the $n_k - 1$-dimensional simplexes or the multidimensional space of frequency vectors. Each elementary step is determined by multiplying the frequency vector by the elementary jump matrix for each coordinate of the frequency vector [2, 3]. The product of the $N^*$-th number of such matrices in the space $R^{N^*}$ determines the transition from the initial vector to the final one in $N^*$ elementary steps.

The set of sequences of elementary matrices from $N^*$ from cofactors is also imposed by the restrictions arising from the condition of the quasi-stationarity of the state of the automaton (Eq. (1)). And following the above assumptions leading to the conditions of the H-theorem, the admissible set of matrices of elementary steps is also limited by the condition of convergence to a quasi-stationary state.

Thus, the interpretation of the model in terms of a random walk on a lattice or in a continuum undergoes significant changes. The meaning of them is as follows: topological restrictions, namely, the nontrivial homotopy type of degeneration space in state and trajectory spaces defines such configurations of states when, defined at the boundary of the some region, they cannot be continued to the interior of the region by continuity, which indicates the presence of discontinuities in attempts of continuous continuation. That is, there is a singularity inside such a region. The kernel of the singularity is a set with partially or completely broken symmetry, i.e., the degeneration space changes with its homotopy type. Thus, in the state spaces and spaces of k-multiple paths [2, 3], defined on these spaces, regions with broken symmetry appear, and these regions have a complex topological nature and are capable of various transformations within their homotopy class.

Examples: Let the space of degeneration is a homogeneous space

$$G/H = SO(3)/SO(2) \times Z_2 = RP^2. \tag{8}$$

$RP^2$ is projective space:

$$\pi_1\left(RP^2\right) = Z_2 \ \ \pi_2\left(RP^2\right) = Z \tag{9}$$

To understand the main points of the model, further examples will be given for low-dimensional spaces of degeneracy. This demonstrates the basic principles of constructing automata that are intuitive in low dimensions. So, the nontriviality of the fundamental group $RP^2$ determines the presence of linear singularities that are closed at their ends by themselves or ending at the boundary of the region. The nontriviality of the second homotopy group, $\pi_2\left(RP^2\right)$of the space of degeneracy, implies the existence of singularities in the spaces that are homotopy equivalent to a point. In the space of paths in the above case, (Eq.(10)) is used to interpret the singularities [12].

$$[W, \Omega X] \leftrightarrow [\Sigma W, X] \tag{10}$$

From which it follows that

$$\pi_i(\Omega X) \cong \pi_{i+1}(X) \tag{11}$$

Consequently,

$$\pi_1(\Omega(G/H)) = \pi_1\left(\Omega\left(SO(3)/SO(2)\times Z_2\right)\right) = \pi_2\left(RP^2\right) = Z \tag{12}$$

According to the above formulas, one-dimensional singularities are not a separately taken state, but a path or loop, which is homotopy equivalent.

The nontriviality of the homotopy groups of the space of degeneracy $G/H$ and the presence of the group structure of homotopy classes give rise to a nontrivial interaction between singularities. In particular, the fusion of singularities is accompanied by the addition of elements of groups of homotopy groups (topological charges) generating singularity data. Other transformations are possible. For example, two linked circular singularities can be continuously transformed into the singularity shown on **Figure 1**.

The interaction of point singularities in the vicinity of a linear singularity can also be nontrivial. In particular, depending on the features of the topological dynamics in the vicinity of the linear singularity, the interacting point singularities may annihilate.

As already noted, the singularity kernel is a region of the state space, the trajectory space, where, in connection with symmetry breaking, trajectories with broken symmetry are realized. In particular, within the framework of the initial space, the degeneracies of the kernel of a singularity represent regions where the state of the automaton is not defined. In this case, the singularity in the original space of degeneracy is interpreted as a discontinuity in the standard topology of $R^{N^*}$, generated by the metric of the Euclidean space. The gap in the trajectory space can be defined in various functional metrics. For example, as shown in **Figure 2**, two trajectories are on opposite sides of the singularity.

For simplicity, let it be a point singularity on a plane. The trajectories shown in **Figure 2** are homotopically non-equivalent. Therefore, an attempt to continuously deform one trajectory into another cannot be continuous. If we are talking about a loop space, then the class of trajectories on both sides of the singularity will undergo the discontinuity in the topology defined in the loop space. That is, the kernel of the singularity contains a class of trajectories, not close to the class of trajectories.

In practice, such trajectories surrounded by quasi-stationary trajectories are already non-quasi-stationary. Removing the degeneracy with respect to the time inversion leads to the appearance of trends in the states of the automaton in such trajectories. The conditions of quasi-stationarity are violated.

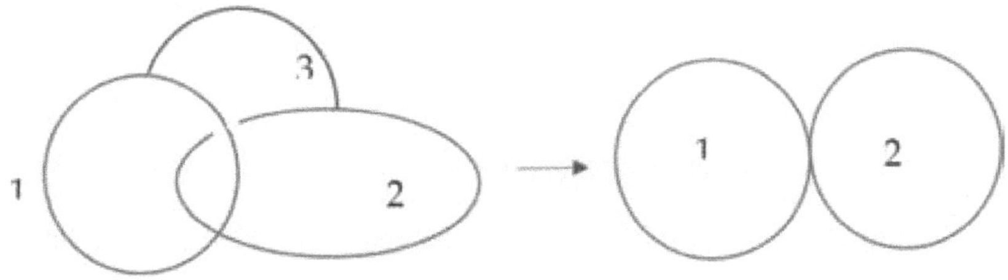


**Figure 1**
*Transformation of two coupled singularities 1 and 2, with singularity 3 connecting them into a bunch of two one-dimensional spheres.*

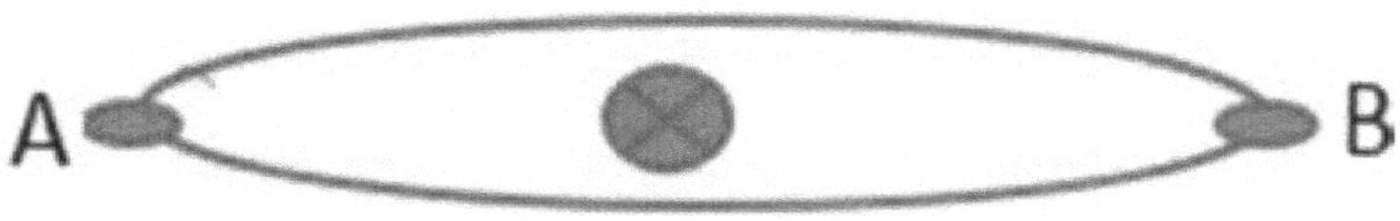


**Figure 2**
*Two non-homotopic trajectories due to a point singularity between them.*

At the same time, the range of permissible sequences in the product of matrices of elementary steps is reduced due to the need to take into account the chronological ordering on the set of products of matrices of elementary transitions. The listed changes do not allow the system to reach some previous states, in view of what new states and new trajectories appear in the system with necessity.

The set of admissible states of automata also changes. In particular, to overcome the prohibitions that have arisen on part of the trajectories, new cells of the automaton are born, that is, an increase in the dimension of the state space and the trajectory space occurs, i.e., there is an increase in the dimension of the state space and the trajectory space. As a result, such change increases the dimension of the symbolic state space represented as $n-1$-dimensional simplex $\Sigma^{n-1}$ . The increase in the simplex dimension and the dimension of the state space generates new types of singularities in the new dimensions and, on the other hand, allows us to again and expand the class of admissible trajectories by increasing the dimension.

In this case, the dimension in the evolution equation changes the conditions of quasi-stationarity and other prohibitions on the set of admissible states expressed as inequalities or equalities also change. An effect similar to the effect of singularity falling into the third dimension in nematic liquid crystals, for example, is observed. Only in the case under consideration is the system's trajectory squeezed onto the verge of a simplex $\Sigma^{K-1}$, $K > n$. The state of the automaton during such a process changes, and the empirical density of the distribution may become non-Gaussian. Thus, with a small movement of the trajectories of the previous class, the extruded path in higher dimensions has a gap in loop space.

Returning to the set of predictive automata, it should be noted that the described scenario of restructuring the internal state of the automaton is far from being the only one, and the topological model offers many different scenarios for the reconstruction of internal states. However, even in the above example, predictors or hidden predictors of reconstructions are identified. Since the task of the prognosis model is to predict exactly the reconstructions, the predictors detected by the automata are predictors of dysfunctions, failure, heart failure, etc. In the example analyzed above, the early predictors are the emergence of the process of birth and destruction of the cells of an automaton; the earlier predictors are associated with the violation of the limiting conditions for the transition probabilities. A change in the stationary conditions leads to an evolution equation that depends on time, respectively; in the state space, there is a change in the transition probabilities over time, which, in turn, determines the trajectories different from the stationary ones. Accordingly, the RUL estimates also change. Returning to the set of predictive automata, it should be noted that the described scenario of restructuring the internal state of the automaton is far from being the only one, and the topological model offers many different scenarios for the reconstruction of internal states. However, even in the above example, predictors or hidden predictors of reconstructions are identified. Since the task of the prognosis model is to predict exactly the reconstructions, the predictors detected by the automata are predictors of dysfunctions, failure, heart failure, etc. In the example analyzed above, the early predictors are the emergence of the process of birth and destruction of the cells of an automaton; the earlier predictors are associated with the violation of the limiting conditions for the transition probabilities. A change in the stationary conditions leads to an evolution equation that depends on time, respectively; in the state space there is a change in the transition probabilities over time, which, in turn, determines the trajectories different from the stationary ones. Accordingly, the RUL estimates also change.

One of the conclusions arising from the symbolic model is associated with the prediction of hidden predictors. The essence of it is as follows. In the process of evolution, any material is subject to degradation. In the materials of which the mechanisms are composed, such as alloys, composite, and polymeric materials, changes occur that affect the functional properties of the material, for example, a local change in the composition of titanium alloys under thermal cycling conditions, grain growth in metals, the formation of dislocations and microcracks, etc. Partially such changes are reflected in the signals of the sensors, if, in addition, structural degradation affects the functional properties of the material. Biomaterials also degrade, for example, the appearance of scar tissue, which replaces the tissue in the myocardium, entailing changes in the ability of the material with the spread of action potential, etc.

With the degradation of the material, the number of class of permissible trajectories changes, as noted above. Taking into account the nontriviality of the homotopy type of degeneration space, cases are allowed when the forbidden trajectory is squeezed onto the verge of a simplex of higher dimension from the class of normal, not forbidden trajectories. Extrusion of a trajectory to higher dimensions leads to changes in the internal state of the automaton. In higher dimensions, there are changes in the number of kinetic or evolution equations, as a result, a change in the conditions of quasi-stationarity and other conditions that limit the set of trajectories, either represented as topological prohibitions, or as equalities or inequalities on the set of admissible states. This leads to the conclusion that in the general case, it is impossible to introduce a metric in the trajectory space in which the development of hidden predictors continuously migrates to the boundary of the heart event or to the boundary of failure. That is, the process of reaching the failure boundary may occur as jump. In this case, the jump is carried out from the class of permissible trajectories, against the background of the absence of dysfunctions by the mechanism of extrusion of the trajectory on the verge of a simplex of higher dimension. In this case, the state of the automaton is transformed in violation of the conditions of quasi-stationarity. Certain analogs of the described effect is noise-induced transitions, but in this case, the system does not require the additional conditions, under which such transitions occur, does not require a Markov property, the fulfillment of the Chapman-Kolmogorov identity, and the conditions imposed on the signal for the validity of the Fokker-Planck equations. In this case, the signal can be both stochastic and chaotic.

In conclusion, it is necessary to stop at the complexity model of the set of automata, begun in [3]. In this case, the interaction between automata is introduced, described in [3] using a metric based on the Radon-Nikodym derivative. However, a simple example should be given of the necessity of complicating automata in order to understand the ultimate goal of such complications. Let us consider an automaton with a two-dimensional cascade or its continual limit. Only such automata are combined into one automaton, for which it can be said that the corresponding stochastic processes are statistically dependent:

$$P(X, Y) - P_1(X)P_2(Y) \neq 0 \tag{13}$$

$P, P_1, P_2$—is the density of distribution function.

We now turn to the degeneration spaces for processes $(X, Y)$. In case of statistical independence

$$G/_H(X, Y) = G/_H(X) \times G/_H(Y) \tag{14}$$

$$\pi_i({}^G\!/_H(X,Y)) = \pi_i\ ({}^G\!/_H(X)) \times \pi_i({}^G\!/_H(Y)) \qquad (15)$$

The situation changes when there is a statistical dependence or some other variant of the interaction of processes. In this case, the formula (13, 14) is unfair. The situation is much more complicated. For example, if Eq. (13) holds true and

$${}^G\!/_H(Y) \subset ({}^G\!/_H(X)) \qquad (16)$$

That homotopy groups can be fundamentally different from the product of the homotopy groups of direct products of spaces of degeneration. In particular, the theorem on the exact sequence of a pair is valid [13]:

$$\ldots \to \pi_i({}^G\!/_H(Y)) \to \pi_i({}^G\!/_H(X)), {}^G\!/_H(Y)) \to \pi_{i-1}({}^G\!/_H(Y)) \to \ldots$$

$$\text{Example} : {}^G\!/_H(X) = S^3, {}^G\!/_H(Y) = S^1$$

$$\ldots \to \pi_2\left(S^3\right) \to \pi_2\left(S^3, S^1\right) \to \pi_{i-1}\left(S^1\right) \to \ldots$$

$$0 \to \pi_2\left(S^3, S^1\ \right) \to \pi_1\left(S^1\right) \to 0$$

$\pi_2\left(S^3, S^1\ \right) = Z$- is the relative homotopy group.

For comparison: $\pi_2\left(S^1\right) = 0$, $\pi_2\left(S^3\right) = 0$.

Thus, in the interaction of automata, singularities appear, described in terms of relative homotopy groups, and these singularities in the general case are fundamentally different from the singularities of the direct product of degeneration spaces. Accordingly, interacting automata exhibit other scenarios of topological dynamics, other predictors and mechanisms for breaking symmetries or lifting degenerations.

The need for complicating the automata described above also determines a number of additional properties of the set of automata, namely, bundle of automata and hypernets.

For a fixed length of the partition of the physically permissible range of values of the sensor readings and in those cases when such partition is determined from the conditions of reducing computational costs or following the traditionally accepted in the general statistics, there is a risk of missing hidden predictors. In this case, bundle of the automaton occurs, which is determined by a shorter partition step. It turns out that each cell of the automaton is also an automaton. This stratification is also necessary in cases where the observed signal has several scale levels.

Formation of hypernets is associated with the formation in the network of interacting automata of a higher dimension. Such an association is not artificial, as was shown in this chapter. When merging automata into the higher-dimensional automaton, the degeneracy space of the finite automaton differs significantly from the degeneration spaces of each of the parent automata.

As a consequence, there is a change in the types of singularities, their topological dynamics, a change in the transformations of the states of automata in the process of their falling out into the faces of the higher dimensions of the simplex, changes in the numerical estimates of the achievement of predicted indicators or limits of failure, heart events, etc.

Ultimately, the architecture and functionality of the predictive monitoring system is as follows. Telemetry enters the computing kernels with automata and preprocessing programs, which includes the wavelet transformation, cleaning the signal by the wavelet filters, and then fed to the recognizing automata. As already noted, the task of predictive automata is reduced to determining the class of the trajectory, determining trends and early predictors based on an analysis of transition probabilities, the number of permissible transition matrices in the product (omega), and also to determine the evolution of this number and the qualitative structure of the set of transition probabilities. It also checks the fulfillment of the stationarity conditions, the behavior of the Kulbak entropy with deviations from the stationarity. Taking into account the revealed evolution of transition probabilities, the trajectory of a walk on the faces of a simplex is determined (predicted). This happens on all the set of automata, including automata with high dimension. Moreover, a causal graph is determined on a subset of high-dimensional automata. Automata that had previously reacted by changing their state to subsequent predictors of later automata are allocated to a separate class. It should be noted here that the selected class is also subject to change, that is, the automata can leave this class or, on the contrary, appear in this class.

The reason for this behavior is as follows:

a. Taking into account the individual characteristics of the mechanism.

b. Individual changes on the sets of transition probabilities and, therefore, on the entire trajectory of states

c. The emergence of new types of singularities due to a change in the structure of the space of degeneration

In a certain class, the earliest automata can be partially placed on the computing power located on the mechanism itself, for example, on the onboard computer, the computing power of the microprocessor of a wearable medical device, etc. Such automata acquire some additional functions; in particular, these automata begin to control the computing kernels of the remote cluster. And, on the contrary, when changing in the whole class, the computing cluster replaces onboard automata with other automata with the same functions. As a result, it becomes possible to optimize the calculations in the system "automata on-board - automata of the computing kernels of remote cluster" up to the advent of opportunities to go offline.

However, periodically online mode is required for inspection. At the same time, the onboard automata connect the online mode based on the determination of their own state, for the most part, with the appearance of birth-death processes, changes in the structure of the set of transition probabilities, the appearance of trends, or rearrangements.

The described situation reflects a more general property of the family of automata, namely, their ability to differentiate. It has already been noted that automata possess the properties of differentiation; a similar property of automata is necessary in cases where an explicit failure predictor is formed. In this case, several automata proceed only to the predictor analysis, for example, when signs of the beginning of a trend in the QT interval are predicted. In this case, besides the set of intervals that capture the cascades corresponding to the QT interval of the ECG, an additional automaton is added to them, processing only the length of the QT interval. The task of the selected family is in this case an estimate of the time to reach critical values of the QT interval. All of the above applies to any traditional geometric characteristics of the ECG as a whole and intervals, complex, and teeth.

## 3. Return on investment

The question about ROI, more precisely, about optimizing the maintenance strategies, was discussed on the basis of the automata model and the hierarchical prognosis model in the works [5, 14]. In the most general formulation, the optimization task is reduced to carrying out the necessary maintenance and variation of the management parameters in order to maximize the stay of the process trajectory in a given homotopy class. The preventive monitoring cluster architecture described above allows reducing the computation time on a remote cluster by transferring a small part of the automata to the monitored object using onboard computing power.

In the field of medical applications (cardiology), we can take as an example the transfer of automaton of the R-R interval to the processor of portable ECG recorder for multi-day monitoring. In addition to this automaton and depending on the general analysis of the evolution of states, automata of a specific purpose are also installed. For example, an automaton for the QT interval, which can autonomously predict a change in the QT interval with an estimate of the time, it takes to reach critical values of a given ECG interval. All of the above is also true for the definition and evolution of such ECG indicators as R-R pause length, QT interval dispersion, and other significant ECG predictors of cardiac events. It should be recalled that the onboard automata can also go offline and manage the computing kernels of the remote cluster. In the above example, such transition is carried out on the basis of the prognosis obtained by the onboard automatic devices. In other cases, an automaton on one of the leads may be sufficient, and a specific cascade corresponding to, for example, the average value of one of the ECG teeth, is monitored.

The described manipulation with automata allows minimizing the cost of the traffic of all ECG leads, as well as the costs associated with the abundance of computations by the entire set of cluster automata.

Of course, the described optimization of calculations increases the comfort of using wearable ECG gadgets, bracelets, and mHealth platforms.

However, the general goal in this case is prognosis of the life-threatening cardiac events, taking preventive measures to eliminate them or minimize the consequences. An example should be the procedure for optimizing drug therapy for atrial fibrillation. The well-known fact that antiarrhythmic drugs used in the treatment of arrhythmias can provoke the appearance of life-threatening arrhythmias. Timely adjustment of the dose of the drug, the rejection of the drug, and its replacement with another drug is one of the actual problems during treatment. In this situation, predictive automata solve the problem of optimizing drug therapy, determining the ineffectiveness of prescribed drugs, thereby reducing the time they are taken. On the other hand, automata predict with an estimate of the probability and time to achieve proarrhythmic effects, which allows the doctor to take steps in advance to correct the treatment.

Consider the field of monitoring engineering for mobile objects, in particular, in transport and directly monitoring for internal combustion engines, hybrid engines. Here, as onboard signaling automata, it offers an automatic device that processes signals from onboard standard sensors, for example, a crankshaft angle sensor or a pressure sensor on common rail systems, etc. With an onboard computer and graphics processors, the number of automata can be increased by covering most of the standard sensors that give an analog or digital signal, for example, the automata processing of the so-called uneven stroke. The evolution of the states of automata that process uneven stroke based on the crankshaft angle sensor reflects the evolution of the cylinder-piston group and, therefore, predicts the development of the most dangerous predictors of failure associated with changes in the combustion mode of the fuel mixture. Adding an accelerometer to the family of standard sensors for analyzing the vibrations of the engine body allows to expand the prediction capabilities for the main friction pairs in the engine mechanics and analyze the status of injectors and high pressure fuel pump (diesel engines). The calculated

RUL estimate allows to calculate an optimizing maintenance strategy in real time. This can also be attributed to railway transport where there are many problems with the analysis of friction pairs, for example, bearings, rotors, etc.

The following fact also contributes to cost reduction and, accordingly, a change in ROI. At the beginning of the operation of automata, the entire set of automata participates in the processing of sensor signals. In the course of time and with the increase in statistical data, the increase in the number of monitored objects of the same type on the set of mechanisms, its clustering becomes possible, that is, the division of mechanisms into classes determined by approximately the same set of predictors necessary for prognosis. In the case of clustering, an additional automaton is introduced, the purpose of which is to determine whether the mechanism belongs to a certain class of the cluster. At the same time, the cumulative chronological database as individual objects or the entire class can significantly reduce the cost of remote computing resource, particularly in cases where onboard computing resource is sufficient for placing on it a few key automata.

The effect of optimization of the computing resource is also provided with the development of self-maintenance or self-recovery systems. As an example of self-recovery systems, there are such failures as coking of nozzles or a piston group of an engine, which is especially common when operating engines subjected to tuning, in violation of the speeds of operation, etc. In such cases, an early prognosis using an automaton on the rotation sensor of the engine and accelerometer allows to determine the initial stages, or rather, the change in the mode of combustion of the fuel mixture at the stage of nucleation of coking centers. However, feeding on several motorcycles of the depleted mixture leads to the dissolution of the germs of growth of the new phase, in this case, the centers of growth of coke deposits. At the same time, substantial savings are quite obvious, since in the more advanced stages of coking, dismantling the nozzle is necessary for its cleaning, processing in an ultrasonic bath, checking at the stand, etc. The same problem is relevant in aviation, where coking of the inner surface of turbine compressor blades is often detected under the conditions of a repair aircraft factory. In this case, expensive blades are simply replaced by new ones. Thus, self-recovery or maintenance of the system or modes can be very effective, especially in the case of early prediction of dysfunctions and at the stage where the group of time inversion enters the degeneration space, which turns the degeneration space into a projective space. In other words, the trajectories in this case are invariant with respect to the time inversion. The presence of such symmetry and, moreover, reversibility is some good approximation of the process at low diffusion velocities in evolution equations of the Fokker-Planck type, for example. It should be noted that the maintenance management task formalized in the previous chapter is most effectively solved exactly in the described class of processes, i.e., in the presence of marked symmetry.

## 4. Conclusions

The prognostic models considered in this chapter are built on the assumption that each observed signal to a certain extent reflects the internal state of the object being studied. In engineering, in medicine, the observed signal is at least indirectly due to the processes occurring in the mechanisms and bioobjects. The task of a complete set of measured parameters as a whole is hardly solvable; one way or another, when studying any object, a model of the object itself is needed, capable of reflecting not only the instantaneous state but also the evolution of the object over time. The observed signal, as a rule, is not obtained directly from an evolving object, because it is impossible to place an accelerometer or a pressure sensor directly into the boiler of a nuclear reactor, glue the patient's heart with electrodes and install a pressure sensor in the working part of the aircraft turbine. Thus, in the end, there is a signal from only ten electrodes from the surface of the human body; the sensors of the machinery are installed far enough from the combustion zone in engines and turbines. Everything briefly listed above leads to the appearance of artifacts of various natures in the signal, and these artifacts must be taken into account in order to isolate the useful part of the signal. In addition to all the above, signals from the so-called distributed systems are usually analyzed. Strong heterogeneity, complex geometry, and complex processes supporting the functionality of these systems are not restored and not reproduced by a set of a dozen sensors. Therefore, by virtue of the above, physical models of objects and their evolution are needed, using the so-called concise description, where, often, an infinite set of measured parameters necessary for modeling the system is replaced by a much smaller number of parameters. The simplest example and classical descriptive compression is the equilibrium thermodynamics, when coordinates and momenta of all particles are reduced in thermodynamic approximation to such quantities as temperature, pressure, etc.

In the cases under consideration, the main emphasis in constructing the prognostic model for evolving distributed systems is to take into account the internal symmetry of the observed signal, without discussing the relationship between the internal signal symmetry and the symmetry inherent in the state of the object and the processes in it.

So, internal symmetry allows representing a signal as a product of slowly varying amplitude and a rapidly oscillating phase cofactor. The phase cofactor has symmetry groups, which acts transitively in pseudo-phase space. Ultimately, the state of the system is described as a vector in multidimensional space, and when taking into account the space of degeneration of a point in the space of degeneration, the evolution of the system is a trajectory of states in the space of degeneration or a multiple space of loops. Further analogies with the theory of topological defects of condensed media make it possible to formalize the evolution of the prediction problem in the form of the evolution of the set of trajectories, their transformations, and a change in the types of trajectories with various kinds of lifting degenerations. The transition to symbolic space, and in fact to the space of probability measures, offers different interpretations of the evolution of a measure in time, in particular, in the form of a complex-arranged walk of the point in a multidimensional simplex. This, in turn, allows to construct families of predictive automata, the evolution of which internal states takes into account the topological nontriviality of the space of degeneracy. Considering the topological nontriviality and, as a result, the topological nontriviality of the set of trajectories determines the complex topological dynamics of the trajectories and the dynamics of the interaction of topological singularities and, most importantly, provides an opportunity for algorithmic construction of the search and definition of predictors of various events associated with changes in the characteristics of the trajectory. Finally, it becomes clear that any change in the characteristic features of the trajectory is associated with the violation of the corresponding symmetry, the removal of the degeneracy in groups or subgroups of the isotropy of the space of degeneracy.

The resulting set of automata has a number of properties that must be mentioned in the context of their further development:

1. Differentiation of automata

2. Bundle of automata

3. Formation of hypernets

Ultimately, the predicting set of automata will have the ability to adapt, i.e., restructuring of its entire structure with changes in the properties of the observed signal, adding signals from new sensors, and changing the operating conditions of engineering.

Thus, the family of automata is capable of analyzing a multitude of signals from various sensors, which makes it possible to use them in all types of engineering and medicine. In the initial stage of operation of the automata, one should limit ourselves to standard onboard diagnostic sensors, which allows one to obtain the necessary data on the need for additional sensors, which, in turn, provides further calculation of the ROI for construction of optimizing maintenance strategies, calculating optimal operating conditions, etc.

## References


[1] Kirillov A, Kirillov S, Pecht M. The Problem of PHM Cloud Cluster in the Context of Development of Self-maintenance and Self-recovery Engineering Systems. In: Proceedings of the 8th World Congress on Engineering Asset Management (WCEAM 2013) & the 3rd International Conference on Utility Management & Safety (ICUMAS); 2013. pp. 1509-1520. DOI: 10.1007/978-3-319-09507-3_128

[2] Kirillov S, Kirillov A, Iakimkin V, Kaganovich Y, Pecht M. Systems of preventive cardiological monitoring: Models, algorithms, first results, perspectives. In: Farhadi H, editor. Medical Internet of Things (m-IoT): Enabling Technologies and Emerging Applications. London: InTechOpen; 2018. ISBN: 978-953-51-6138-7. In press

[3] Kirillov S, Kirillov A, Iakimkin V, Pecht M. Remote computing cluster for optimization of preventive maintenance strategies: models, algorithms. In: Márquez FPG, editor. Maintenance Management. London: InTechOpen; 2018. ISBN: 978-953-51-8088-3. In press

[4] Kirillov S, Kirillov A, Iakimkin V, Khodos F, Kaganovich Y, Pecht M. PHM applications in medicine and medical implantable device. In: Proceedings of the IEEE Prognostics and System Health Management Conference (PHM-Chengdu); 2016. DOI: 10.1109/PHM.2016.7819755

[5] Kirillov A, Kirillov S, Pecht M. The calculating PHM cluster: CH&P mathematical models and algorithms of early prognosis of failure. In: Proceedings of the Prognostics and System Health Management Conference (PHM–Beijing); 2012. DOI: 10.1109/PHM.2012.6228771

[6] Iakimkin V, Kirillov A, Kirillov S. Distributed Pre-processing of telemetry for mobile engineering objects. In: Proceedings of the 8th World Congress on Engineering Asset Management (WCEAM 2013) & the 3rd International Conference on Utility Management & Safety (ICUMAS). pp. 463-472. DOI: 10.1007/978-3-319-09507-3_40

[7] Ziman J. Model of Disorder. Cambridge: Cambridge University Press; 1979

[8] De Gennes PG. Scaling Concepts in Polymer Physics. New York: Cornell University Press; 1979

[9] Mermin ND. Topologies theory of defects. Reviews of Modern Physics. 1979;**51**(3):591-648

[10] Iakimkin V, Kirillov A, Kirillov S. Preprocessing of the observed data and the recognition of the hidden signs of fault. In: Proceedings of the 4th IEEE Conference on Prognostics and System Health Management (PHM); 2013. Vol. 33. DOI: 10.3303/CET1333012

[11] Hida T. Brownian Motion. New York: Springer-Verlag; 2012

[12] Adams J. Infinite Loop Spaces. Princeton: Princeton University Press; 1978

[13] Fuchs D, Fomenko A, Gutenmacher V. Homotopic Topology. Budapest: Akadémiai Kiadó; 1986

[14] Kirillov A, Kirillov S, Pecht M. Remote calculating PHM cluster: The first results. In: Proceedings of the 4th IEEE Conference on Prognostics and System Health Management (PHM); 2013. Vol. 33. DOI: 10.3303/CET1333045

Chapter 4

# Fault Diagnosis Techniques for a Wind Turbine System


## Abstract

The fault diagnosis and prognosis of wind turbine systems represent a challenging issue, thus justifying the research topics developed in this work with application to safety-critical systems. Therefore, this chapter addresses these research issues and demonstrates viable techniques of fault diagnosis and condition monitoring. To this aim, the design of the so-called fault detector relies on its estimate, which involves data-driven methods, as they result effective methods for managing partial information of the system dynamics, together with errors, model-reality mismatch and disturbance effects. In particular, the considered data-driven strategies use fuzzy systems and neural networks, which are employed to establish non-linear dynamic links between measurements and faults. The selected prototypes are based on non-linear autoregressive with exogenous input descriptions, since they are able to approximate non-linear dynamic functions with arbitrary degree of accuracy. The capabilities of the designed fault diagnosis schemes are verified via a high-fidelity simulator, which describes the normal and the faulty behaviour of a wind turbine plant. Finally, the robustness and the reliability features of the proposed methods are validated in the presence of uncertainty and disturbance implemented in the wind turbine simulator.



**Keywords:** fault diagnosis, analytical redundancy, fuzzy prototypes, neural networks, diagnostic residuals, fault reconstruction, wind turbine simulator


## 1. Introduction

The increasing level of wind-generated energy in power generation worldwide also increases the levels of reliability and the so-called 'sustainability' shown by wind turbines. Wind turbine systems should generate the required amount of electrical power continuously, depending on the available wind speed, the grid's demand and possible malfunctions.

To achieve this aim, possible malfunctions affecting the process have to be properly detected and managed, before they degrade the nominal working conditions of the plant or become critical issues. Wind turbines with large rotors (i.e., of megawatt size) are very expensive systems, thus requiring an extremely high level of availability and reliability, in order to maximise the generated energy (at a reduced cost), with a minimisation of the operation and maintenance (O&M) services. In fact, the costs of the produced energy are mainly due to the installation cost of the wind turbine, while unplanned O&M costs could increase it up to about 30%, in particular when offshore installations are considered, see Odgaard [1].

These issues have motivated the development of fault diagnosis techniques that can be coupled with the fault-tolerant controllers (the so-called 'sustainable' systems). On the other hand, many turbine manufacturers adopt conservative approaches against faults, which lead to the shutdown of the plant in order to wait for O&M service. Hence, effective tools for coping with faults have to be investigated, in order to improve wind turbine features, particularly during faulty situations. This will lead to prevent critical failures that may affect other wind turbine components, thus avoiding unplanned replacement of functional parts, as well as the decrease of O&M costs, with the increase of the energy production. Moreover, the development of digital control systems, big data tools and artificial intelligence strategies enhance the development of new real-time condition monitoring, diagnosis and fault-tolerant control strategies for industrial processes, which can be available only on demand.

In recent years, many works have been proposed on the topics of fault diagnosis of wind turbines, as shown very recently in Habibi et al. [2] and Lan et al. [3]. Some of them are focused on the diagnosis of particular faults, for example, those affecting the drive-train system at a wind turbine level. Sometimes, these faults are better managed when the wind turbine system is considered in comparison to other parts of the whole plant, see Odgaard & Stoustrup [4]. Moreover, fault-tolerant control of wind turbines has been investigated, for example, in Parker et al. [5] and international cooperations on these problems were also proposed, see Odgaaard and Shafiei [6].

Fault diagnosis oriented to the sustainability feature when applied to safety-critical systems such as wind turbines has been proven to be a challenging issue, see Byrski and Byrski [7] and Xu et al. [8], thus motivating the research topics addressed in this chapter.

This point is fundamental as the increasing demand for energy generation using renewable sources has led to higher attention on renewable energy conversion systems, and in particular wind turbines. They represent very complex and safety-critical plants which require reliability, availability, maintainability and safety. Moreover, their efficiency to generate electrical power has to be maximised. This motivates novel research aspects, in particular in the context of diagnosis and control. The earlier diagnosis of faults and sustainable control solutions can lead to optimise energy conversion and guarantee the desired performances in the presence of possible malfunctions due to unexpected faults and disturbance.

Therefore, this chapter analyses the problem of the fault diagnosis for wind turbine systems, and the development of practical and reliable solutions for fault diagnosis, also known as fault detection and isolation (FDI). Further design of fault-tolerant controllers is not considered in this work, but it can rely on the tools considered in this chapter. In fact, the fault diagnosis module provides information on the faulty or fault-free conditions of the system, so that the controller activity can be compensated. This fault diagnosis task is enhanced by the use of fault estimators, which are obtained via data-driven approaches, as they offer effective tools for managing limited analytical knowledge of the process dynamics, together with noise and disturbance effects.

The first data-driven solution considered in this chapter uses fuzzy Takagi-Sugeno models, see Babuška [9], which are derived from a clustering algorithm, followed by an identification procedure, see Simani et al. [10]. A second solution is also considered, which relies on neural networks to describe the non-linear analytical links between measurement and fault signals. The chosen network architecture belongs to the Nonlinear AutoRegressive with eXogenous (NARX) input prototype, which can describe dynamic relationships along time. The training of the neural network fault estimators exploits standard training algorithm, that processes the data acquired from the process, see Roy and Ganguli [11].

The developed fault diagnosis strategies are verified by means of a high-fidelity simulator, which describes the normal and the faulty behaviour of a wind turbine plant. The achieved performances are verified in the presence of uncertainty and disturbance effects, thus validating the robustness features of the proposed schemes. The effectiveness verified from the achieved results suggests further investigations on more realistic applications of the proposed schemes.

The work is organised as follows. Section 2 recalls the wind turbine simulator. Section 3 illustrates the fault diagnosis methodologies relying on fuzzy and neural network prototypes. The obtained results are summarised in Section 4. Finally, Section 5 ends the chapter by outlining the key achievements of the study, and providing suggestions for future research issues.

## 2. Wind turbine simulator description

The wind turbine simulator used in this work was proposed in Odgaard et al. [12]. It describes the realistic behaviour of a three-blade horizontal-axis variable-speed pitch-controlled wind turbine coupled with a full converter generator. The overall system consists of four interconnected modules, that is, the wind driving process, the wind turbine, the measurement system and the baseline controller. The wind turbine block contains three submodels: the blade and the pitch system, the drive-train model and the generator system. The links between the system submodels are represented in **Figure 1**, with the fault diagnosis system to be designed. The simulator is able to generate several fault scenarios, see Odgaard et al. [12].

In the following, the description of these interconnected submodels is briefly recalled.

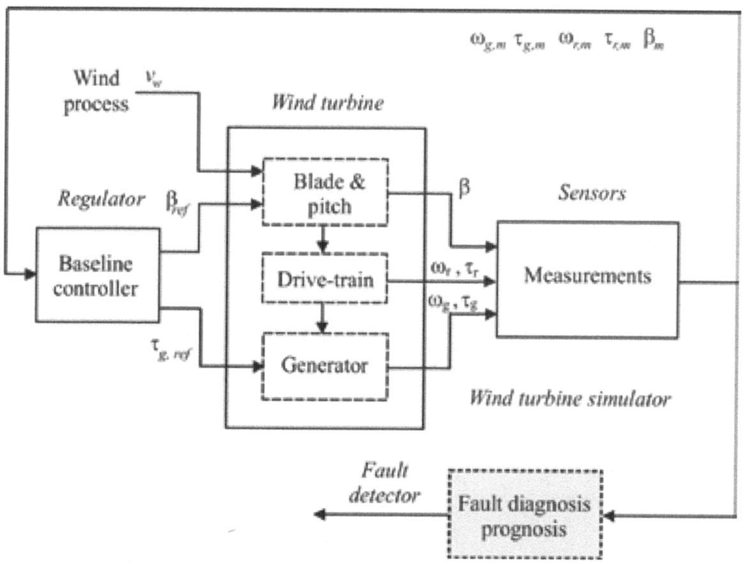


**Figure 1.**
*The wind turbine simulator with its fault diagnosis system.*

## 2.1 Wind turbine model

The turbine system consists of three submodels motivated by the power transmission flow. First, the blade and pitch block represents how the blades capture wind energy, which is based on the following aerodynamic law:

$$\tau_r(t) = \frac{\rho \pi R^3 C_q(\lambda(t), \beta(t)) v_w^2(t)}{2} \tag{1}$$

For each blade, Eq. (1) describes the torque acting on the rotor $\tau_r$, depending on the squared wind speed $v_w^2$, the air density $\rho$ and the rotor radius $R$. The coefficient $C_q$ is usually defined using a two-dimensional map depending on the blade pitch angle $\beta$ and the tip-speed ratio $\lambda$, that is, the ratio between the linear velocity of the blade tip and the wind speed. This map is represented by means of a look-up table. The blade and pitch system includes the dynamics of the pitch angle hydraulic piston servo system, which is approximated as a second-order transfer function of Eq. (2):

$$\frac{\beta(s)}{\beta_{ref}(s)} = \frac{\omega_n^2}{s^2 + 2\zeta\omega_n s + \omega_n^2} \tag{2}$$

where $\beta_{ref}$ is the reference pitch angle computed by the turbine controller, while $\zeta$ and $\omega_n$ are the transfer function parameters.

The drive-train system determines the power flow through the gear box from the rotor toward the electric generator, whose dynamics are described as in Eq. (3):

$$\begin{cases} J_r \dot{\omega}_r = \tau_r - K_{dt}\theta_\Delta - (B_{dt} + B_r)\omega_r + \dfrac{B_{dt}}{N_g}\omega_g \\ J_g \dot{\omega}_g = \dfrac{\eta_{dt} K_{dt}}{N_g}\theta_\Delta + \dfrac{\eta_{dt} B_{dt}}{N_g}\omega_r - \left(\dfrac{\eta_{dt} B_{dt}}{N_g^2} + B_g\right)\omega_g - \tau_g \\ \dot{\theta}_\Delta = \omega_r - \dfrac{\omega_g}{N_g} \end{cases} \tag{3}$$

where $J_r$ and $J_g$ are the inertia moments of the rotor and generator shafts, respectively. $K_{dt}$ is the torsion stiffness, $B_{dt}$ is the torsion damping factor, $B_g$ is the viscous friction of the generator shaft, $B_r$ is the viscous friction of the low-speed shaft, $N_g$ is the gear ratio, $\eta_{dt}$ is the efficiency and $\theta_\Delta$ is the torsion angle.

Finally, the generator submodel represents the converter dynamics by means of first-order transfer function of Eq. (4):

$$\frac{\tau_g(s)}{\tau_{g,ref}(s)} = \frac{\alpha_g}{s + \alpha_g} \tag{4}$$

where $\tau_{g,ref}$ is the reference torque defined by the controller and $\alpha_g$ is the transfer function parameter.

Finally, the generated power $P_g$ is computed as the product of the generator torque by its speed, decreased by the efficiency coefficient $\eta_g$:

$$P_g = \eta_g \omega_g \tau_g \tag{5}$$

As sketched in **Figure 1**, the signals generated by the wind turbine system are assumed to be acquired through the measurement block, whose objective is to simulate the real behaviour of sensors and actuators. Therefore, the measured signals are modelled as sum of their actual value and white Gaussian process terms. Moreover, the wind turbine simulator includes a baseline controller, represented by a PID standard regulator, which regulates the generated power on the basis of the actual wind speed, as shown in Odgaard & Stoustrup [4] and Odgaard et al. [12].

### 2.2 Simulated fault scenario

The wind turbine simulator includes the generation of three different typical fault cases, that is, sensor, actuator and system faults, see Odgaard and Stoustrup [4] and Odgaard et al. [12].

For the case of the sensor faults, they are generated as additive signals on the affected measurements. As an example, the faulty sensor of faulty pitch angle $\beta_m$ provides wrong measurements on blade orientation; thus, if not handled, the controller cannot fully track the power reference signal.

On the other hand, actuator faults lead to the alteration of pitch angle or the generator torque transfer functions of Eqs. (2) and (4), by modifying their dynamics. They simulate a pressure drop in the hydraulic circuit of the pitch actuator or an electronic break down in the converter device.

Finally, a system fault affects the drive train of the turbine, which is described as a slow variation in time of the friction coefficient. This can be due to the effect of wear and tear along time of the mechanical parts.

These nine fault cases are summarised in **Table 1**, which also highlights which measured signals are affected by them, as shown in **Figure 1**.

With these assumptions, the overall model of the wind turbine process can be represented as a non-linear continuous-time function $\mathbf{f}_{wt}$ describing the evolution of the turbine state vector $\mathbf{x}_{wt}$ excited by the input vector $\mathbf{u}$:

$$\begin{cases} \dot{\mathbf{x}}_{wt}(t) = \mathbf{f}_{wt}(\mathbf{x}_{wt}, \mathbf{u}(t)) \\ \mathbf{y}(t) = \mathbf{x}_{wt}(t) \end{cases} \tag{6}$$

where in this case, the state of the system is considered equal to the monitored system output, that is, the rotor speed, the generator speed and the generated power:

$$\mathbf{x}_{wt}(t) = \mathbf{y}(t) = \left[\omega_{g,m1}, \omega_{g,m2}, \omega_{r,m1}, \omega_{r,m2}, P_{g,m}\right]$$

| Fault case | Fault type | Affected measurement |
|---|---|---|
| 1 | Sensor | $\beta_{1,m1}$ |
| 2 | Sensor | $\beta_{2,m2}$ |
| 3 | Sensor | $\beta_{3,m1}$ |
| 4 | Sensor | $\omega_{r,m1}$ |
| 5 | Sensor | $\omega_{r,m2}$ and $\omega_{g,m2}$ |
| 6 | Actuator | Pitch system of blade #2 |
| 7 | Actuator | Pitch system of blade #3 |
| 8 | Actuator | $\tau_{g,m}$ |
| 9 | System | Drive train |

**Table 1.**
*Wind turbine simulator fault scenario.*

On the other hand, the input vector:

$$\mathbf{u}(t) = \left[\beta_{1,m1}, \beta_{1,m2}, \beta_{2,m1}, \beta_{2,m2}, \beta_{3,m1}, \beta_{3,m2}, \tau_{g,m}\right]$$

consists of the measurements of the pitch angles from the three redundant sensors, as well as the measured torque. These signals are sampled with sample time $T$ in order to acquire a number $N$ of data $\mathbf{u}(k)$, $\mathbf{y}(k)$ with $k = 1, \ldots, N$, in order to implement the data-driven fault diagnosis solutions proposed in this chapter.

## 3. Fault diagnosis techniques

This chapter considers two data-driven approaches, relying on fuzzy system and neural network structures, which are used to design the fault diagnosis schemes. Therefore, this section briefly introduces the general scheme of the fault diagnosis strategy, by recalling the basic features of the fuzzy systems and neural networks, as addressed in Sections 3.1 and 3.2, respectively. Moreover, these architectures, which are represented by NARX structures, are exploited residual generators for solving the problem of fault diagnosis, according to the analytical redundancy principle, see Chen and Patton [13].

In order to solve the fault diagnosis problem, this work assumes that the wind turbine system is affected by *equivalent* additive faults on the input and the output measurements, as well as measurement errors, as described by Eq. (7):

$$\begin{cases} \mathbf{u}(k) = \mathbf{u}^*(k) + \tilde{\mathbf{u}}(k) + \mathbf{f}_u(k) \\ \mathbf{y}(k) = \mathbf{y}^*(k) + \tilde{\mathbf{y}}(k) + \mathbf{f}_y(k) \end{cases} \tag{7}$$

where $\mathbf{u}^*(k)$ and $\mathbf{y}^*(k)$ represent the actual process variables, $\mathbf{u}(k)$ and $\mathbf{y}(k)$ are the measurements acquired from the sensors, while $\tilde{\mathbf{u}}(k)$ and $\tilde{\mathbf{y}}(k)$ describe the measurement errors. According to the description of Eq. (7), signals of the faults $\mathbf{f}_u(k)$ and $\mathbf{f}_y(k)$ also have *equivalent* additive effects. Obviously, these functions are different from zero in faulty cases. In general, the vector $\mathbf{u}(k)$ has $r$ components, that is, the number of process inputs, while $\mathbf{y}(k)$ has $m$ elements, that is, the number of process outputs.

Among the possible approaches exploited for residual generation, and based on the analytical redundancy principle, this work proposes to exploit fuzzy system and neural network structures, which provide an on-line estimation $\hat{\mathbf{f}}(k)$ of the fault signals $\mathbf{f}_u(k)$ and $\mathbf{f}_y(k)$. Hence, as shown in **Figure 1**, the so-called diagnostic residuals $\mathbf{r}(k)$ are equal to the estimated fault signals, $\hat{\mathbf{f}}(k)$, which are computed by the general fault estimator, as highlighted by Eq. (8):

$$\mathbf{r}(k) = \hat{\mathbf{f}}(k) \tag{8}$$

The variable $\hat{\mathbf{f}}(k)$ is the generic fault vector, that is, $\hat{\mathbf{f}}(k) = \left\{\hat{f}_1(k), \ldots, \hat{f}_{r+m}(k)\right\}$. Therefore, the general fault estimate $\hat{f}_i(k)$ can be equal to one of the $i$ components of the fault vectors $\mathbf{f}_u(k)$ or $\mathbf{f}_y(k)$ in Eqs. (7), with $i = 1, \ldots, r+m$.

The residual generation scheme exploiting the fault estimators as residual generator is depicted in **Figure 2**. Note that this strategy is able to provide both the fault detection and isolation tasks, that is, the fault diagnosis function, see Chen and Patton [13].

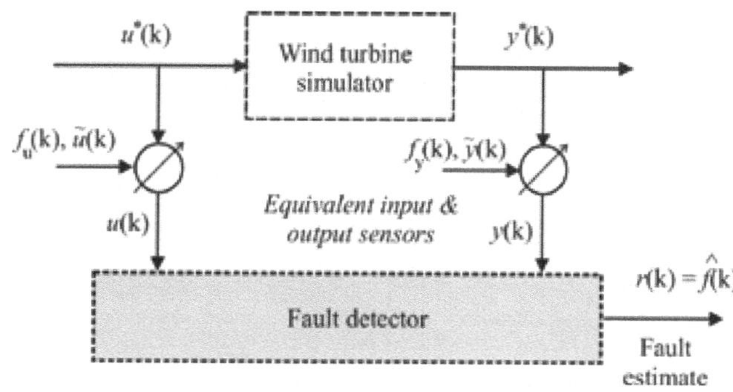


**Figure 2.**
*Fault detectors for fault diagnosis.*

**Figure 2** shows that in general the residual generators use the acquired input and output measurements $\mathbf{u}(k)$ and $\mathbf{y}(k)$. As first step, the fault diagnosis scheme consists of the fault detection task. In this case, as the residual is equal to the estimated fault signal, it is easily performed via a proper thresholding logic directly operating on the residual itself, without requiring complex elaboration with proper evaluation functions, as shown in Chen and Patton [13]. Therefore, the occurrence of the $i$th fault can be simply detected via the threshold logic of Eq. (9) applied to the $i$th residual $r_i(k)$:

$$\begin{cases} \overline{r}_i - \delta\sigma_{r_i} \leq r_i \leq \overline{r}_i + \delta\sigma_i & \text{fault-free case} \\ r_i < \overline{r}_i - \delta\sigma_{r_i} \text{or } r_i > \overline{r}_i + \delta\sigma_{r_i} & \text{faulty case} \end{cases} \tag{9}$$

with $r_i(k)$ representing the $i$th component of the vector $\mathbf{r}(k)$. If it is considered as a random variable, its means $\overline{r}_i$ and variance $\sigma^2_{r_i}$ values can be estimated in fault-free condition, after the acquisition of $N$ samples, according to Eq. (10):

$$\begin{cases} \overline{r}_i = \dfrac{1}{N}\sum\limits_{k=1}^{N} r_i(k) \\ \sigma^2_{r_i} = \dfrac{1}{N}\sum\limits_{k=1}^{N} \left(r_i(k) - \overline{r}_i\right)^2 \end{cases} \tag{10}$$

Note that the parameter $\delta \geq 2$ represents a tolerance variable, which has to be properly tuned in order to effectively separate the fault-free from the faulty conditions. A common choice of $\delta$ can rely on the three-sigma rule, otherwise extensive simulations can be exploited for optimising this $\delta$ value, see Chen and Patton [13].

Once the fault detection phase is accomplished, the fault isolation task is directly obtained by means of a bank of estimators. As described by Eq. (7), the faults are considered as equivalent signals that are injected and affect the input measurements via the signal $\mathbf{f}_u$, or the output measurements by means of $\mathbf{f}_y$.

According to the scheme depicted in **Figure 3**, in order to uniquely isolate one of the input or output faults, under the assumption that multiple faults cannot occur, a bank of multi-input single-output (MISO) fault estimators is designed. In general, the number of this estimators is equal to the number of faults that have to be diagnosed, that is, which coincides with the number of input and output measurements, $r + m$. Therefore, the $i$th estimator providing the reconstruction of the fault $\hat{f}(k) = r_i(k)$ is driven by the components of the input and output signals $\mathbf{u}(k)$ and $\mathbf{y}(k)$, respectively. These components are selected in order to be sensitive to the

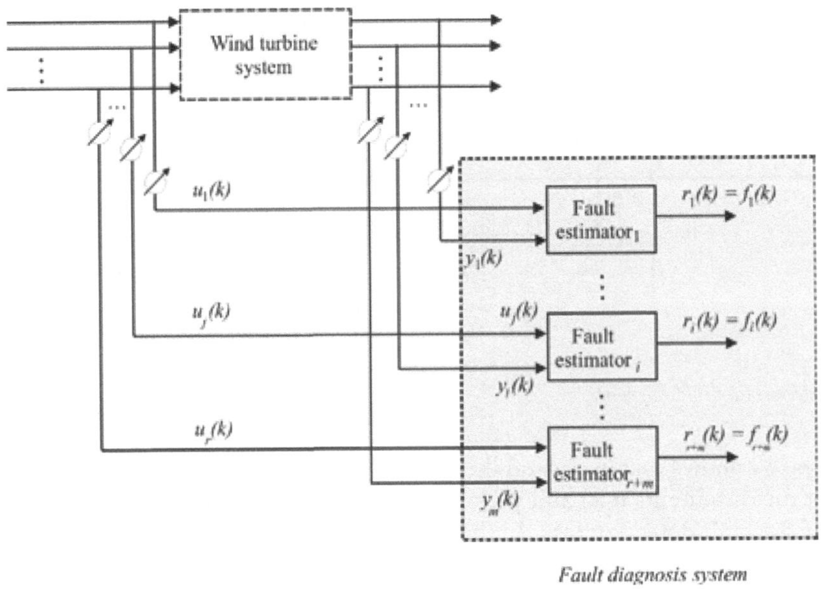


**Figure 3.**
*Residual generators bank with $r_i(k) = f_i(k)$.*

specific fault $f_i(k)$. In fact, the design of these fault estimators is enhanced by the fault sensitivity analysis described in Section 3.3. For each case, the fault modes and their resulting effects on the rest of the system are analysed, and in particular the most sensitive input $u_j(k)$ and output $y_l(k)$ measurements to that specific fault situation are selected. In this way, by means of the fuzzy system and neural network tools, it will be possible to derive the dynamic relationships between the input-output measurements, $u_j(k)$ and $y_l(k)$, and the faults $f_i(k)$, as highlighted by **Figure 3**.

**Figure 3** shows this fault estimator bank, where the fault estimators are driven by the input-output signals selected via the fault sensitivity analysis procedure. In this way, the residual $r_i(k) = \hat{f}_i(k)$ is insensitive only to the fault affecting those inputs and outputs, $u_j(k)$ and $y_l(k)$, defined by the selector blocks. It is worth noting that, using this configuration, multiple faults occurring at the same time cannot be correctly isolated.

As already remarked, the sensitivity analysis, which has to be executed before the design of the fault estimators, suggests how to select the input-output signals feeding the fault estimator modules. After this selection procedure is performed, as described in Section 3.3, the design of the fuzzy or neural network models is achieved, as recalled in Sections 3.1 and 3.2, respectively. Finally, the threshold test logic of Eq. (9) allows the achievement of the fault diagnosis task.

### 3.1 Fuzzy system modelling and identification

This section describes the design of the fault estimators described by means of the Takagi-Sugeno (TS) prototypes, see Takagi and Sugeno [14]. Therefore, the unknown relationships between the selected measurements and the faults are described by fuzzy models, which consist of a number of rules. These rules connect the measured signals acquired from the system under diagnosis to its faults, described in form of IF⇒THEN relations, processed by a fuzzy inference system (FIS), see Babuška [9].

According to this approach, the approximation of non-linear multi-input single-output (MISO) systems can be achieved by the Takagi-Sugeno (TS) fuzzy reasoning, as described in Babuška [9]. The TS modelling approach proposed here, as addressed in Takagi and Sugeno [14], describes the consequents as deterministic functions $g_i(\cdot)$ of the inputs, while the antecedents remain fuzzy propositions.

The fuzzy rule of the FIS has the form of Eq. (11):

$$R_i : IF \quad (\text{fuzzy combination of inputs}) \quad THEN \quad \text{output} = g_i(\text{inputs}) \tag{11}$$

where $i$ refers to the number of rules. The antecedents are combined by means of membership functions $\lambda_i(\mathbf{x})$ that take into account the logical connectives expressed by linguistic propositions. The rule consequent function $g_i(\cdot)$ is defined as parametric function in the affine form of Eq. (12):

$$g_i(\mathbf{x}) = \mathbf{a}_i^T \mathbf{x} + b_i \tag{12}$$

where $\mathbf{a}_i$ is the parameter vector, and $b_i$ is a scalar offset, while $g_i(\mathbf{x})$ is the $i$th rule output. The number of rules is supposed equal the number of clusters $n_C$ used for partitioning the data into regions where the relations $g_i(\cdot)$ hold, see Babuška [9]. Furthermore, the antecedent of each rule defines the degree of fulfilment for the corresponding consequent model, defined by the membership function $\lambda_i(\mathbf{x})$. Therefore, the global model is expressed as a fuzzy composition of parametric models $g_i(\mathbf{x})$.

The TS prototype takes the form of the expression of Eq. (13):

$$\hat{f} = \frac{\sum_{i=1}^{n_C} \lambda_i(\mathbf{x}) g_i(\mathbf{x})}{\sum_{i=1}^{n_C} \lambda_i(\mathbf{x})} \tag{13}$$

Using this fuzzy approach, in general, the fault $\hat{f}$ can be reconstructed from suitable data acquired from the system under diagnosis. In other words, the fault $\hat{f}$ is a weighted average of affine functions $g_i(\mathbf{x})$ of the input-output measurements, where the weights are the combined degree of fulfilment $\lambda_i(\mathbf{x})$ of the system inputs.

It is worth noting that the system under investigation corresponds to the wind turbine process described in Section 2, which has a dynamic behaviour. Therefore, the considered input vector $\mathbf{x}$ of the TS model of Eq. (13) contains the current as well as delayed samples of the system input and output signals.

Therefore, in order to include dynamics into the static relation of Eq. (11), the consequents are described as discrete-time linear AutoRegressive models with eXogenous input (ARX) of order $o$, in which the regressor vector has the form of Eq. (14):

$$\mathbf{x}(k) = \left[\ldots, y_l(k-1), \ldots, y_l(k-o), \ldots u_j(k), \ldots, u_j(k-o), \ldots\right]^T \tag{14}$$

where $u_l(\cdot)$ and $y_j(\cdot)$ are the components of the actual system input and output vectors $\mathbf{u}(k)$ and $\mathbf{y}(k)$ is selected via the fault sensitivity analysis tool of Section 3.3, and exploited in the scheme of **Figure 3**. The variable $k$ represents the time step, with $k = 1, 2, \ldots, N$. The affine parameters associated to the $i$th model of the Eq. (12) are collected into the vector:

$$\mathbf{a}_i = \left[\alpha_1^{(i)}, \ldots, \alpha_o^{(i)}, \delta_1^{(i)}, \ldots, \delta_o^{(i)}\right]^T \tag{15}$$

where the $\alpha_j^{(i)}$ coefficients refer to the output samples, while $\delta_j^{(i)}$ are associated to the input ones.

A powerful approach to the design of the $i$th FIS as approximator for the system under diagnosis begins with the partitioning of the available data $\mathbf{u}(k)$ and $\mathbf{y}(k)$ of Eq. (7) into subsets, known as cluster. A cluster is defined as a set of data that are more similar to each other rather than to the members of another cluster. The similarity among data can be expressed in terms of their distance from a particular item, exploited as the cluster prototype. Fuzzy clustering provides an effective tool to obtain a partitioning of data in which the transitions among subsets are smooth, rather than abrupt. Moreover, fuzzy clustering assumes that the data of each cluster are characterised by an affine behaviour, which is indeed modelled by the relation of Eq. (12). Different clustering methods have been proposed in literature, see for example, more recent works Graaff and Engelbrecht [15] and Jun et al. [16].

With reference to this work, the design of the FIS is considered as a system identification problem from the noisy data of Eqs. (7). In fact, the estimation of the consequent parameters $\mathbf{a}_i$ and $b_i$ of Eq. (12) is required using the input-output data for designing the bank of the fault estimations reported in **Figure 3**. Moreover, the data are acquired from the measurements selected from the procedure suggested in Section 3.3. The identification scheme exploited in this work was proposed by the authors in Fantuzzi et al. [17]. This approach is based on the minimisation of the prediction errors of the individual TS local affine models considered as $n_C$-independent estimation problems. Their solutions rely on the estimation of errors-in-variables models in Fantuzzi et al. [17], which is also the assumption represented by Eq. (7).

Another key aspect, which is not considered here, regards the determination of the optimal number of clusters $n_C$, as the clustering algorithm assumes that the number of clusters $n_C$ has been fixed. These issues are considered in the development of the estimation procedure properly integrated by the authors, which also determines the antecedent degrees of fulfilment $\mu_{ik}$ required by Eq. (13) and solved with curve fitting methods, see Babuška [9].

### 3.2 Neural network modelling and training

This study proposes a different data-driven approach, based on neural networks, which is exploited to implement the fault diagnosis block. This section briefly recalls their general structure and properties, which are used to implement the fault estimators.

Therefore, according to the scheme shown in **Figure 4**, a bank of neural networks is realised in order to reproduce the behaviour of the faults affecting the system under diagnosis using a proper set of input and output measurements. The neural network structure consists of different layers of neurons, also known as *perceptron,* see Haykin [18], modelled as a static function $f$. This function is described by an *activation function* with multiple inputs properly weighted by unknown parameters that determine the learning capabilities of the whole network.

A categorisation of these learning structures concerns the way in which their neurons are connected to each other, see Xu et al. [19]. This work proposes to use a feedforward network, also called multilayer perceptron, where the neurons are grouped into unidirectional layers. The first of them, the input layer, is directly fed by the network inputs; then, a hidden layer takes the inputs from the neurons of the input layer and transmits them the output to the neurons of the third layer, the output layer, which produces the final network outputs. According to this structure, neurons are connected from one layer to the next, but not within the same layer. The only constraint is the number of neurons in the output layer, that has to be equal to the number of actual network outputs. On the other hand, recurrent

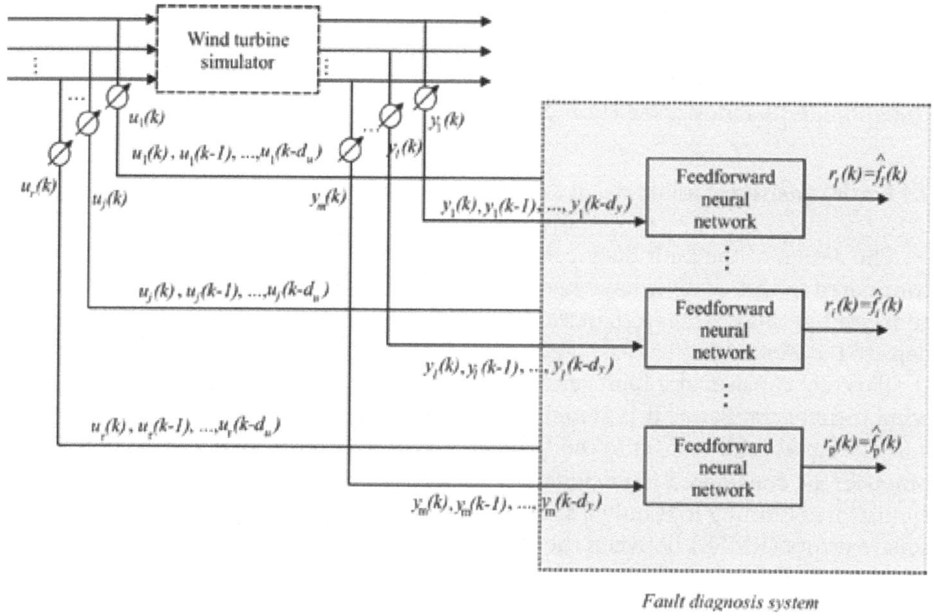


**Figure 4.**
*Neural networks as fault estimators with* $\hat{f}_i(k) = r_i(k)$.

networks are multilayer networks, in which the output of some neurons is fed back to neurons belonging to previous layers, thus the information flow in forward as well as in backward directions, allowing a dynamic memory inside the network, see Hunt et al. [20].

A noteworthy intermediate solution is provided by the multilayer perceptron with a tapped delay line, which is a feedforward network whose inputs come from a delay line. This study proposes to use this solution, defined as quasistatic neural network, as it represents a suitable tool to predict dynamic relationships between the input-output measurements and the considered fault functions. In this way, another NARX description is obtained, since the non-linear (static) network is fed by the delayed samples of the system inputs and outputs selected by the fault sensitivity analysis tool described in Section 3.3. Indeed, if properly trained, the NARX network can estimate the current (and the next) fault samples $f_j(k)$ on the basis of the selected past measurements of system inputs and outputs $u_l(k)$ and $y_j(k)$, respectively, in the same way of the fuzzy systems.

Therefore, with reference to the $i$th residual generator of **Figure 4**, which is used to design the estimator bank of **Figure 3**, this NARX network is described by the relation of Eq. (16):

$$\hat{f}_i(k) = F(\ldots, u_j(k), \ldots, u_j(k-d_u), \ldots y_l(k-1), \ldots, y_l(k-d_y), \ldots) \tag{16}$$

where $\hat{f}_i(k)$ is the estimation of the generic $i$th fault, while $u_j(\cdot)$ and $y_l(\cdot)$ are the generic $j$th and $l$th components of the measured inputs and outputs $\mathbf{u}$ and $\mathbf{y}$, respectively, that are selected via the fault sensitivity analysis tool. $k$ is the time step, $d_u$ and $d_y$ are the number of delay of inputs and outputs, respectively, which have to be properly selected. $F(\cdot)$ is the function realised by the static neural network, which depends on the layer architecture, the number of neurons, their weights and their activation functions. The NARX network used as generic fault $f_i(k)$ estimator is depicted in **Figure 4**.

The design parameters are represented by the number of neurons and the number of delays of the network inputs and outputs, while the value of the weights of each neuron are derived from the network training from the data acquired from the system under diagnosis, see Hunt et al. [20].

### 3.3 Fault sensitivity analysis

The design of the fault diagnosis schemes proposed for the application example considered in this chapter have been summarised in Section 4. However, the tool addressed in this chapter enhances the design of the banks of these fault estimators depicted in **Figure 3**.

This tool consists of a fault sensitivity analysis that has to be performed on the wind turbine simulator. It is aimed at defining the most sensitive measurements $u_j(k)$ and $y_l(k)$ with respect to the fault conditions $f_i(k)$ considered in Section 2.2. In practice, the considered fault signals have been injected into the wind turbine simulator, assuming that only a single fault may occur. Then, the relative mean square errors (RMSE) between the fault-free and faulty measured signals are evaluated, so that, for each fault, the most sensitive signal $u_j(k)$ and $y_l(k)$ can be selected. The results of the fault sensitivity analysis are summarised in **Table 2** for the wind turbine system.

In particular, the fault sensitivity analysis is conducted on the basis of a selection algorithm that is performed by introducing the normalised sensitivity function $N_x$, defined in Eq. 17:

$$N_x = \frac{S_x}{S_x^*} \tag{17}$$

with

$$S_x = \frac{\|x_f(k) - x_n(k)\|_2}{\|x_n(k)\|_2} \tag{18}$$

and

$$S_x^* = \max \frac{\|x_f(k) - x_n(k)\|_2}{\|x_n(k)\|_2} \tag{19}$$

The value of $N_x$ indicates the effect of the considered fault case with respect to the general measured signal $x(k)$, with $k = 1, 2, ..., N$. The subscripts 'f' and 'n' indicate the faulty and the fault-free case, respectively. Therefore, the

| **Fault $f_i$** | 1 | 2 | 3 | 4 | 5 |
|---|---|---|---|---|---|
| **Measurements $u_j, y_l$** | $\beta_{1,m1}$ | $\beta_{2,m2}$ | $\beta_{3,m1}$ | $\omega_{r,m1}$ | $\omega_{r,m1}$ |
| **RMSE** | 11.29 | 0.98 | 2.48 | 1.44 | 1.45 |
| **Fault $f_i$** | 6 | 7 | 8 | 9 | |
| **Measurements $u_j$ or $y_l$** | $\beta_{2,m1}$ | $\beta_{3,m2}$ | $\tau_{g,m}$ | $\omega_{g,m1}$ | |
| **RMSE** | 0.80 | 0.73 | 0.84 | 0.77 | |

**Table 2.**
*Fault sensitivity $f_i(k)$ with respect $u_j(k)$ and $y_l(k)$.*

| Fault case $f_i$ | Most sensitive inputs $u_j$ | Most sensitive outputs $y_l$ |
|---|---|---|
| 1 | $\beta_{1,m1}, \beta_{1,m2}$ | $\omega_{g,m2}$ |
| 2 | $\beta_{1,m2}, \beta_{2,m2}$ | $\omega_{g,m2}$ |
| 3 | $\beta_{1,m2}, \beta_{3,m1}$ | $\omega_{g,m2}$ |
| 4 | $\beta_{1,m2}$ | $\omega_{g,m2}, \omega_{r,m1}$ |
| 5 | $\beta_{1,m2}$ | $\omega_{g,m2}, \omega_{r,m2}$ |
| 6 | $\beta_{1,m2}, \beta_{2,m1}$ | $\omega_{g,m2}$ |
| 7 | $\beta_{1,m2}, \beta_{3,m2}$ | $\omega_{g,m2}$ |
| 8 | $\beta_{1,m2}, \tau_{g,m}$ | $\omega_{g,m2}$ |
| 9 | $\beta_{1,m2}$ | $\omega_{g,m1}, \omega_{g,m2}$ |

**Table 3.**
*Fault sensitivity test.*

measurements that are most affected by the considered fault lead to a value of $N_x$ equal to 1. Otherwise, a smaller value of $N_x$, that is, close to zero, represents a signal $x(k)$ not affected by the fault. Those signals characterised by high value of $N_x$ are thus selected as the most sensitive measurements, and they will be considered in the design of the fault diagnosis modules of the bank sketched in **Figure 3**.

The complete results of the fault sensitivity analysis are summarised in **Table 3**. For each fault case, the selected signals of the wind turbine benchmark are marked as inputs or outputs.

This method represents a key feature of the proposed approach to fault diagnosis. In fact, the fault estimators of the bank of **Figure 3** can be designed by exploiting a reduced number of signals, thus leading to a noteworthy simplification of the overall complexity, and a decrease in the computational cost of the training and identification phases.

## 4. Simulation results

This section summarises the simulations performed with the considered wind turbine benchmark, and the performances of the proposed fault diagnosis solutions. Due to the presence of the uncertainty and disturbance effects included in the benchmark, the robustness features of the developed fault diagnosis techniques are also verified in simulation.

With reference to the wind turbine benchmark of Section 2, all simulations are driven by the same wind mean speed sequence. It was acquired from a real measurement of wind speed, which represents a good coverage of typical operating conditions, as it ranges from 5 to 20 m/s, with a few spikes at 25 m/s, see Odgaard et al. [12]. The simulations last for 4400 s, with single fault occurrences. The discrete-time simulator runs at a sampling frequency of 100 Hz, so that $N = 440{,}000$ samples are acquired during each simulation. With reference to the different fault cases reported in Section 2.2, **Table 4** shows the shape and the timing of the fault modes affecting the process. They model input (actuator) or output (sensor) additive faults, which are used for sensitivity analysis of Section 3.3.

As an example, in order to highlight the actual fault effect on the wind turbine measurements, **Figure 5** shows the fault sensitivity test. In particular, the cases of the faults 1, 2, 3 and 8 in fault-free and faulty conditions are depicted.

| Fault case | Fault type | Fault shape | Occurrence (s) |
|---|---|---|---|
| 1 | Actuator | Step | 2000–2100 |
| 2 | Actuator | Step | 2300–2400 |
| 3 | Actuator | Step | 2600–2700 |
| 4 | Actuator | Step | 1500–1600 |
| 5 | Actuator | Step | 1000–1100 |
| 6 | Sensor | Step | 2900–3000 |
| 7 | Sensor | Trapezoidal | 3500–3600 |
| 8 | Sensor | Step | 3800–3900 |
| 9 | Sensor | Step | 4100–4300 |

**Table 4.**
*Wind turbine simulator fault conditions.*

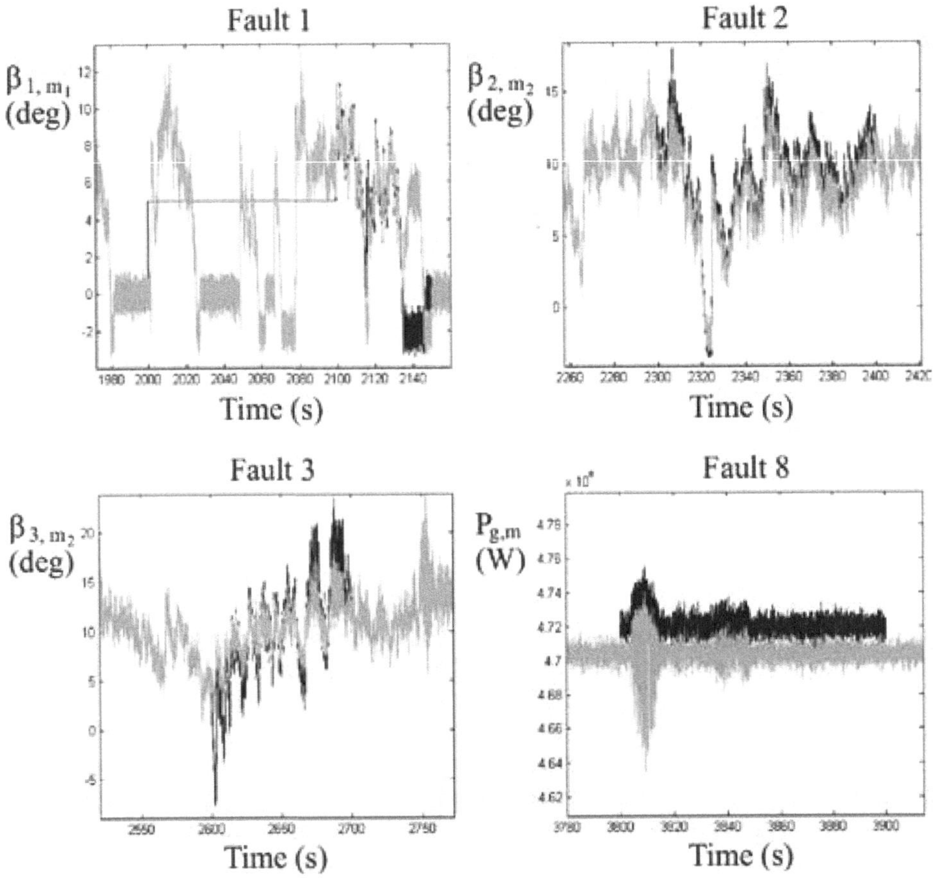


**Figure 5.**
*Example of fault-free (grey line) and faulty (black line) signals.*

## 4.1 Fuzzy estimators for fault diagnosis

The problem of the fault diagnosis of the wind turbine simulator is solved in this work by designing fuzzy prototypes as fault reconstructors. The considered approach is different from the one presented in Simani et al. [21], where the fuzzy models were used as output predictors.

Section 3.1 suggested to exploit the fuzzy $c$-means clustering algorithm. When applied to the data of the wind turbine simulator, a number $n_C = 4$ of clusters and $o = 3$ delays on input and output regressors were determined. The tool also generated the membership function points that are fitted through Gaussian membership functions. After data clustering, the regressands $\alpha_j^{(i)}$ and $\delta_j^{(i)}$ of Eq. (15) were identified for each cluster by following the procedure of Section 3.1. The TS models of Eq. (13) were thus implemented and nine fault estimators were designed, built and organised into the estimator scheme in order to accomplish the fault diagnosis task, as sketched in **Figure 3**.

The effectiveness of the fuzzy TS fault estimators used was assessed in terms of root mean squared error (RMSE), which is computed as the difference between the predicted $\hat{f}_i(k)$ and the actual fault $f_i(k)$ signals for each of the fuzzy estimators, with $i = 1, \dots, 9$. **Table 5** summarises the achieved performance of the nine fault estimators of **Figure 3**.

In this case, these estimated signals $\hat{f}_i$ are directly exploited as diagnostic residuals $r_i$, as remarked by Eq. (8). They can be compared with the thresholds of Eq. (9), optimally selected in order to achieve the optimisation of the overall fault diagnosis performance indices, in terms of missed fault and the false alarm rates, see Ding [22]. In particular, **Table 6** summarises the values of the parameter $\delta$ of Eq. (9) for each fault estimator $i$.

Note that, in general, each of the nine fuzzy fault estimators described by the relations of Eqs. (13) and (14) has three inputs (see **Table 3**), with a number of delays $n = 3$ and $n_C = 4$ clusters. Therefore, the number of estimated parameters for each fuzzy MISO model (three inputs and one output) is equal to $(3 + 1) \times n = 12$. Moreover, for each fault estimator, the estimation of the fuzzy membership functions $\lambda_i(\cdot)$ of Eq. (13) with $i = 1, \dots, n_C$ was required.

In the following, the main simulation results are summarised. Two actuator faults $f_u$ and two sensor fault $f_y$ are considered, namely the fault cases 1, 4, 8 and 9 of the scenarios recalled in Section 2.2.

According to **Table 3**, these faults caused the alteration of the monitored input and output signal **u**, **y** affecting the residual $r_1 = \hat{f}_1$, $r_4 = \hat{f}_4$, $r_8 = \hat{f}_8$ and $r_9 = \hat{f}_9$ generated by the fuzzy fault estimators. These faults $\hat{f}_i$ depicted in **Figure 6** demonstrate

| **Fault estimator $\hat{f}_i$** | 1 | 2 | 3 | 4 | 5 |
|---|---|---|---|---|---|
| RMSE | 0.016 | 0.023 | 0.021 | 0.020 | 0.019 |
| **Fault estimator $\hat{f}_i$** | 6 | 7 | 8 | 9 | |
| RMSE | 0.021 | 0.017 | 0.021 | 0.019 | |

**Table 5.**
*Fuzzy fault estimator capabilities with RMSE.*

| **Residual $r_i(k)$** | 1 | 2 | 3 | 4 | 5 | 6 | 7 | 8 | 9 |
|---|---|---|---|---|---|---|---|---|---|
| $\delta$ | 3.8 | 4.3 | 4.2 | 4.5 | 3.7 | 4.4 | 4.3 | 3.5 | 3.9 |

**Table 6.**
*The parameter $\delta$ for the threshold selection.*

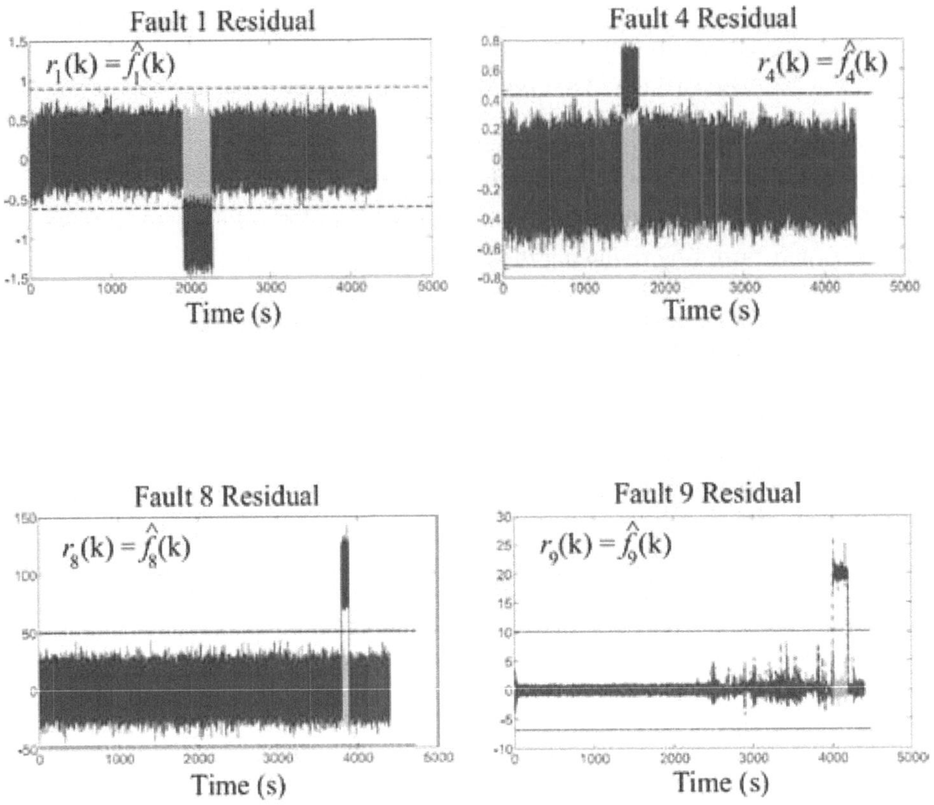


**Figure 6.**
*Fault-free (grey line) and faulty (black continuous line) residuals with faults 1, 4, 8 and 9.*

the achievement of the fault diagnosis task, as they exceed the threshold levels only when the relative fault is active, as recalled in **Table 4**.

**Figure 6** depicts the reconstructed fault functions $\hat{f}_i(k)$ generated by the fuzzy estimators in faulty conditions (black continuous line) with respect to the fault-free residuals (grey line). The fixed thresholds are depicted with dotted lines. The considered residuals refer to the fault cases 1, 4, 8 and 9. It is worth noting that in fault-free conditions, the estimated fault functions $\hat{f}_i(k)$ are not zero due to both the model-reality mismatch. **Figure 6** also highlights the robustness and reliability features of the developed fuzzy estimators.

### 4.2 Neural networks for fault diagnosis

As for the fuzzy systems, nine NARX neural networks described in Section 3.2 were designed to estimate the nine faults affecting the acquired measurements, according to the scheme of **Figure 3**. The neural networks selected for fault diagnosis purpose consist of 3 layers, with 3 neurons in the input layer, 16 in the hidden one, and 1 neuron in the output layer. A number of $d_u = d_y = 4$ delays were selected in the relation of Eq. (16). Both the input and the hidden layers used sigmoidal activation functions, while the output layer exploits the linear one. According to **Table 3** and **Figure 4**, each of the nine neural networks was driven by three inputs.

As for the fuzzy models, the prediction efficacy of the designed neural networks was verified in terms of RMSE. The achieved results are summarised in **Table 7**, which were obtained by comparing the estimated faults with respect to the simulated ones.

| Fault estimate $\hat{f}_i(k)$ | 1 | 2 | 3 | 4 | 5 |
|---|---|---|---|---|---|
| RMSE | 0.009 | 0.009 | 0.009 | 0.012 | 0.011 |
| **Fault estimate $\hat{f}_i(k)$** | 6 | 7 | 8 | 9 | |
| RMSE | 0.011 | 0.009 | 0.009 | 0.014 | |

**Table 7.**
*Neural network performances.*

| $r_i(k)$ | 1 | 2 | 3 | 4 | 5 | 6 | 7 | 8 | 9 |
|---|---|---|---|---|---|---|---|---|---|
| $\delta$ | 4.2 | 4.9 | 4.7 | 5.1 | 4.2 | 4.6 | 4.8 | 4.1 | 4.3 |

**Table 8.**
*$\delta$ values for the threshold selector.*

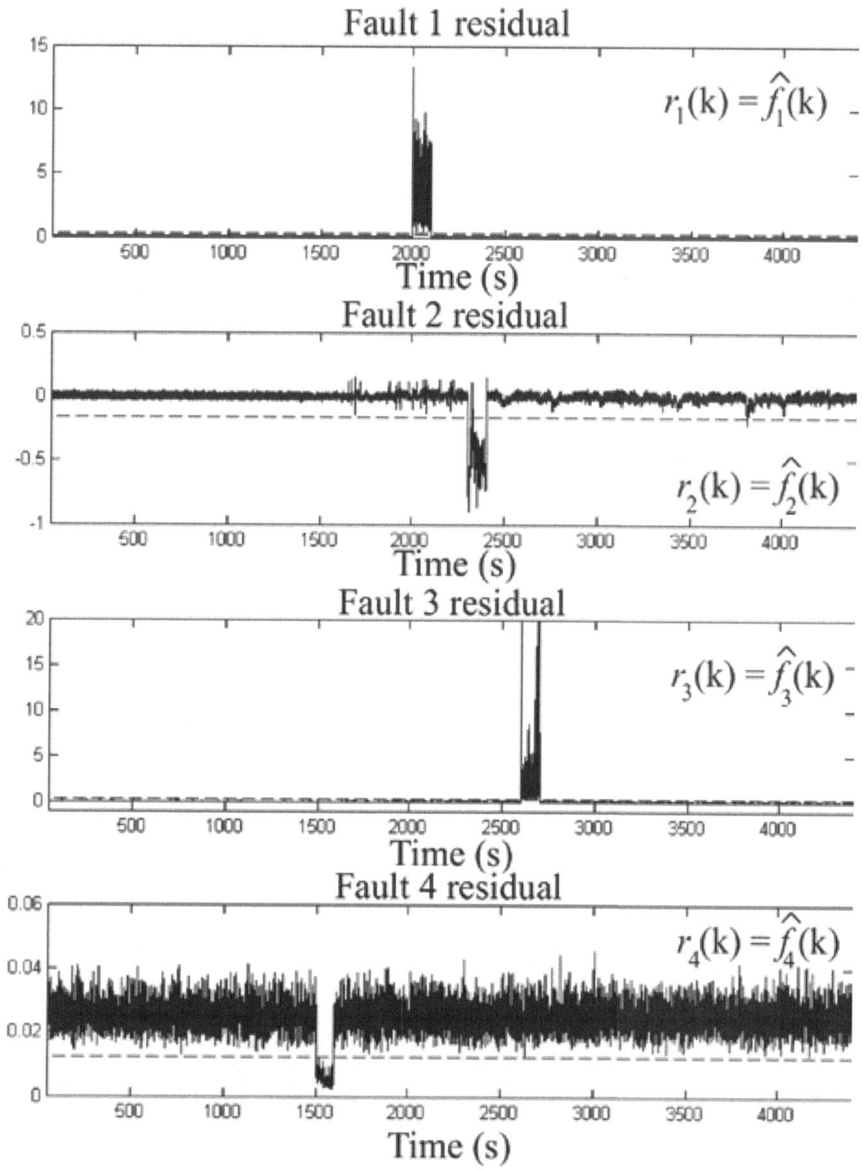


**Figure 7.**
*Estimated faults (continuous line) $\hat{f}_i(k)$ and thresholds (dashed line) for cases 1, 2, 3 and 4.*

The fault diagnosis task is thus achieved by comparing the residuals $r_i = \hat{f}_i(k)$ of Eq. (8) with fixed optimised thresholds, as described by Eq. (9). As for the fuzzy estimators, the values of the parameter $\delta$ of Eq. (9) for each fault estimator $i$ is summarised in **Table 8**.

On the other hand, **Figure 7** shows an example of residual signals for the fault cases 1, 2, 3 and 4, together with the selected thresholds.

In particular, **Figure 6** depicts the residuals $\hat{f}_i(k)$ generated in faulty conditions by the neural network estimators (continuous line) compared with the fixed thresholds (dashed line). The considered residuals refer to the faults $f_1(k), f_2(k)$, $f_3(k)$ and $f_4(k)$ of **Table 4**.

The achieved results show the effectiveness of the proposed fault diagnosis solutions, also with respect to disturbance and uncertainty effects on the wind turbine simulator, thus highlighting their potential application to real wind turbine systems.

## 5. Conclusion

The chapter studied data-driven tools for solving the problem of the fault diagnosis and prognosis of a wind turbine process. The design of this fault detector is based on the estimate of the fault itself, achieved by means of artificial intelligence methods. They were considered since these viable tools demonstrated to be able to cope with poor information on the process dynamics, in the presence of errors, model-reality mismatch and disturbance effects. In particular, these methodologies rely on fuzzy and neural network structures used to determine the non-linear dynamic links between measurements and fault signals. The selected structures belong to the non-linear autoregressive with exogenous input architectures, since they may model any non-linear dynamic relationship with arbitrary degree of accuracy. The fault diagnosis and prognosis strategies were validated via a high-fidelity simulator of a wind turbine process. The achieved performances in terms of reliability and robustness were thus tested by considering the presence of uncertainty and disturbance effects modelled by this wind turbine simulator. Further works will verify the features of the same fault diagnosis schemes when applied to real plants.

## References

[1] Odgaard PF. FDI/FTC wind turbine benchmark modelling. In: Patton RJ, editor. Workshop on Sustainable Control of Offshore Wind Turbines. Vol. 1. Hull, UK: Centre for Adaptive Science & Sustainability, University of Hull; 2012

[2] Habibi H, Nohooji HR, Howard I. Adaptive PID control of wind turbines for power regulation with unknown control direction and actuator faults. IEEE Access. 2018;**6**: 37464-37479. DOI: 10.1109/ACCESS.2018.2853090

[3] Lan J, Patton RJ, Zhu X. Fault–tolerant wind turbine pitch control using adaptive sliding mode estimation. Renewable Energy. 2018;**116**(Part B): 219-231. DOI: 10.1016/j.renene.2016.12.005

[4] Odgaard PF, Stoustrup J. A benchmark evaluation of fault tolerant wind turbine control concepts. IEEE Transactions on Control Systems Technology. 2015;**23**(3):1221-1228

[5] Parker MA, Chong HN, Ran L. Fault-tolerant control for a modular generator–converter scheme for direct–drive wind turbines. IEEE Transactions on Industrial Electronics. 2011;**58**(1): 305-315

[6] Odgaaard PF, Shafiei SE. Evaluation of wind farm controller based fault detection and isolation. In: Proceedings of the IFAC SAFEPROCESS Symposium 2015. Vol. 48. IFAC; Paris, France: Elsevier; 2015. pp. 1084-1089. DOI: 10.1016/j.ifacol.2015.09.671

[7] Byrski J, Byrski W. A double window state observer for detection and isolation of abrupt changes in parameters. International Journal of Applied Mathematics and Computer Science. 2016;**3**(26):585-602. DOI: 10.1515/amcs-2016-0041

[8] Xu F, Puig V, Ocampo-Martinez C, Olaru S, Niculescu SI. Robust MPC for actuator–fault tolerance using set-based passive fault detection and active fault isolation. International Journal of Applied Mathematics and Computer Science. 2017;**27**(1):43-61. DOI: 10.1515/amcs-2017-0004

[9] Babuška R. Fuzzy Modeling for Control. Boston, USA: Kluwer Academic Publishers; 1998

[10] Simani S, Fantuzzi C, Rovatti R, Beghelli S. Parameter identification for piecewise linear fuzzy models in noisy environment. International Journal of Approximate Reasoning. 1999;**1**(22): 149-167. Elsevier

[11] Roy N, Ganguli R. Filter design using radial basis function neural network and genetic algorithm for improved operational health monitoring. Applied Soft Computing Journal. 2006;**6**(2):154-169

[12] Odgaard PF, Stoustrup J, Kinnaert M. Fault–tolerant control of wind turbines: A benchmark model. IEEE Transactions on Control Systems Technology. 2013;**21**(4):1168-1182. ISSN: 1063–6536. DOI: 10.1109/TCST.2013.2259235

[13] Chen J, Patton RJ. Robust Model–Based Fault Diagnosis for Dynamic Systems. Boston, MA, USA: Kluwer Academic Publishers; 1999

[14] Takagi T, Sugeno M. Fuzzy identification of systems and its application to modeling and control. IEEE Transactions on Systems, Man, and Cybernetics. 1985;**SMC-15**(1): 116-132

[15] Graaff AJ, Engelbrecht AP. Clustering data in stationary environments with a local network neighbourhood artificial immune

system. International Journal of Machine Learning and Cybernetics. 2012;**3**(1):1-26. DOI: 10.1007/s13042-011-0041-0

[16] Jun W, Shitong W, Chung F-L. Positive and negative fuzzy rule system, extreme learning machine and image classification. International Journal of Machine Learning and Cybernetics. 2011;**2**(4):261-271. DOI: 10.1007/s13042-011-0024-1

[17] Fantuzzi C, Simani S, Beghelli S, Rovatti R. Identification of piecewise affine models in noisy environment. International Journal of Control. 2002; **75**(18):1472-1485. DOI: 10.1109/87.865858

[18] Haykin S. Kalman Filtering and Neural Networks, Adaptive and Learning Systems for Signal Processing, Communications, and Control. New York, USA: Wiley–Interscience, John Wiley & Sons, Inc; 2001

[19] Xu J-X, Liu C, Hang C. Combined adaptive and fuzzy control using multiple models. In: Proceedings of Third IEEE International Conference on Fuzzy Systems, Orlando, FL. 1994

[20] Hunt K, Sbarbaro D, Zbikowki R, Gawthrop P. Neural networks for control system: A survey. IEEE Transactions on Neural Networks. 1992; **28**:1083-1112

[21] Simani S, Farsoni S, Castaldi P. Fault diagnosis of a wind turbine benchmark via identified fuzzy models. IEEE Transactions on Industrial Electronics. 2015;**62**(6):3775-3782. Invited paper for the special issue "Real–time fault diagnosis and fault tolerant control". DOI: 10.1109/TIE.2014.2364548

[22] Ding SX. Model–Based Fault Diagnosis Techniques: Design Schemes, Algorithms, and Tools. 1st ed. Berlin Heidelberg: Springer; 2008. ISBN: 978–3540763031

Chapter 5

# Analytic Prognostic in the Linear Damage Case Applied to Buried Petrochemical Pipelines and the Complex Probability Paradigm


## Abstract

In 1933, *Andrey Nikolaevich Kolmogorov* established the system of five axioms that define the concept of mathematical probability. This system can be developed to include the set of imaginary numbers by adding a supplementary three original axioms. Therefore, any experiment can be performed in the set $\mathcal{C}$ of complex probabilities which is the summation of the set $\mathcal{R}$ of real probabilities and the set $\mathcal{M}$ of imaginary probabilities. The purpose here is to include additional imaginary dimensions to the experiment taking place in the "real" laboratory in $\mathcal{R}$ and hence to evaluate all the probabilities. Consequently, the probability in the entire set $\mathcal{C} = \mathcal{R} + \mathcal{M}$ is permanently equal to one no matter what the stochastic distribution of the input random variable in $\mathcal{R}$ is; therefore the outcome of the probabilistic experiment in $\mathcal{C}$ can be determined perfectly. This is due to the fact that the probability in $\mathcal{C}$ is calculated after subtracting from the degree of our knowledge the chaotic factor of the random experiment. Consequently, the purpose in this chapter is to join my complex probability paradigm to the analytic prognostic of buried petrochemical pipelines in the case of linear damage accumulation. Accordingly, after the calculation of the novel prognostic model parameters, we will be able to evaluate the degree of knowledge, the magnitude of the chaotic factor, the complex probability, the probabilities of the system failure and survival, and the probability of the remaining useful lifetime; after that a pressure time $t$ has been applied to the pipeline, which are all functions of the system degradation subject to random and stochastic influences.



**Keywords:** probability norm, complex probability set, degree of our knowledge, chaotic factor, remaining useful lifetime, degradation, analytic prognostic, linear damage


## 1. Introduction

> *"An intellect which at any given moment knew all the forces that animate Nature and the mutual positions of the beings that comprise it, if this intellect were vast enough to submit its data to analysis, could condense into a single formula the movement of the greatest bodies of the universe and that of the lightest atom: for such*

> *intellect nothing could be uncertain; and the future just like the past would be present before its eyes".*
>
> *Marquis Pierre-Simon de Laplace.*

> *"The Divine Spirit found a sublime outlet in that wonder of analysis, that portent of the ideal world, that amphibian between being and not-being, which we call the imaginary root of negative unity".*
>
> *Gottfried Wilhelm von Leibniz.*

The high availability of technological systems, like defense, aerospace, automobile industries, and petrochemistry, is a central major objective of previous and latest developments in the technology of system design. Pipelines are the primary component of the systems of hydrocarbon transport in petrochemical industries. They are vital for human activities because they serve to transport water, natural gases, and oil from sources to all consumer sites. A novel analytic prognostic model was established in my earlier research work and applied to the case of pipelines subject to the effects of corrosion, to soil loading, and to internal pressure. These will initiate micro-cracks in the body of the tubes that can spread suddenly and can lead to failure. The increase of pipeline availability and the reduction of their global mission cost and performance necessitate to elaborate a suitable process of prognostic. Accordingly, a novel strategy based on degradation analytic laws was applied to diverse dynamic systems and was developed in my research work [1–6]. Additionally, the remaining useful lifetime (*RUL*) was predicted and calculated from a predefined threshold of degradation. Based on a system of a physical petrochemical pipeline, my publications developed a strategy to design a model of failure prognostic that will be more elaborated and further enhanced in the present book chapter.

Moreover, prognostic is a process involving a prediction capacity. Using prognostic, we are able to evaluate the equipment remaining useful lifetime in terms of its future usage and its history of functioning. Predicting the remaining useful lifetime of industrial systems turns out to be presently a vital goal for industrialists knowing that the consequences of failure, which can occur suddenly, are usually very expensive. The traditional maintenance strategies [7, 8] founded on a static threshold of alarm are no more practical and efficient since they do not consider the instantaneous functioning state of a product. The establishment of a prognostic approach as an "intelligent" maintenance consists of the health follow-up, monitoring, and analysis, based on physical measurements utilizing sensors.

Also, earlier expert studies of prognostic belong in general to three categories of technical approaches: the first category is the "experience-based prognostic" [9] which is based on measurements taken from a machine health monitoring, for example, those based on stochastic model, expert judgment, Bayesian approach, reliability analysis, Markovian process, optimization of preventive maintenance, etc. Their methodology of prognostic shows to be simple but inflexible toward changes in the environment and in the system behavior. The second category is the "estimation-based or trending prognostic" based on the statistics of vast measured data. We can cite as illustrations the work relying on the behavior of degradation expressed by abaci and utilizing a system expert description (process-mission-environment) [10]; the work relying on artificial intelligence, machine learning [11], neural network [12], and fuzzy logic [13]; and additionally the work based on dissipativity-based fuzzy integral sliding mode control of continuous time T-S fuzzy systems, SMC design for robust stabilization of nonlinear Markovian jump singular systems, sliding mode control of fuzzy singularly perturbed systems with application to electric circuits, the stabilization of quantized sampled-data neural network-based control systems, etc. Their methodologies are designated generally as not very precise, but they propose a powerful tool to the theory of prognostic. The third category is the "model-based prognostic" relying on the mathematical description of the process of degradation and its evolution level utilizing nondestructive inspection (NDI) monitoring. It is designated to be more precise and flexible than the two first categories. My earlier research illustrates a methodology of analytical prognostic relying on analytic laws of damage, such as the linear damage accumulation law of Palmgren-Miner and the fatigue crack propagation law of Paris-Erdogan. It fits in the third category of models. This approach is used whenever the law of damage of the studied system is analytically available. The advantage of this approach is consequently its precise and realistic features in evaluating the remaining useful lifetime of a system [14–17].

Additionally, pipes are petrochemical systems that transport natural gas and oil in huge quantities and over long distances. Their life prognostic is crucial in this industry because their availability has vital outcomes. Their major failures are due to soil settlements, seismic ground waves, deformations, buckling, internal and external corrosion, vibration and resonance, stress concentration in welding and fitting, and pressure fluctuation over long period. The failures due to fatigue by means of cracks propagation are noticed and measured by the tools of crack detection. Therefore, three case studies of pipelines were taken into consideration in my earlier publications [18, 19]: buried, unburied, and subsea (offshore pipelines). Each one of these situations necessitates different physical parameters like friction and soil pressure, atmospheric and water pressure, and corrosion. The buried pipes case will only be considered in the present chapter.

## 2. The purpose and the advantages of the present work

Computing probabilities is the main work of classical probability theory. Adding new dimensions to the stochastic experiments will lead to a deterministic expression of probability theory. This is the original idea at the foundations of this work. Actually, the theory of probability is a nondeterministic system in its essence; that means that the event outcomes are due to the chance and randomness. The addition of novel imaginary dimensions to the chaotic experiment occurring in the set $\mathcal{R}$ will yield a deterministic experiment, and hence a stochastic event will have a certain result in the complex probability set $\mathcal{C}$. If the random event becomes completely predictable, then we will be fully knowledgeable to predict the outcome of stochastic experiments that arise in the real world in all stochastic processes. Consequently, the work that has been accomplished here was to extend the real probabilities set $\mathcal{R}$ to the deterministic complex probabilities set $\mathcal{C} = \mathcal{R} + \mathcal{M}$ by including the contributions of the set $\mathcal{M}$ which is the imaginary set of probabilities. Therefore, since this extension was found to be successful, then a novel paradigm of stochastic sciences and prognostic was laid down in which all stochastic phenomena in $\mathcal{R}$ was expressed deterministically. I called this original model "the Complex Probability Paradigm" that was initiated and illustrated in my 12 research publications. [20–31].

Furthermore, although the analytic linear prognostic laws are deterministic and very well-known in [14, 16], there are chaotic and stochastic influences and aspects (such as humidity, temperature, material nature, geometry dimensions, applied load location, water action, corrosion, soil pressure and friction, atmospheric pressure, etc.) that influence the buried pipeline system and make its function of degradation diverge from its computed trajectory modeled by these deterministic laws. An updated follow-up of the degradation performance and behavior with cycle number or time, which is subject to non-chaotic and chaotic influences, is made possible by what I called the system failure probability due to its definition that estimates the jumps in the function of degradation $D$.

Additionally, my objective in this present work is to connect the complex probability paradigm to the buried pipeline system analytic prognostic in the case of linear damage accumulation which is subject to fatigue. In fact, the system failure probability derived from prognostic will be applied to and included in the complex probability paradigm. This will lead to the original and novel model of prognostic illustrated in this chapter. Thus, by determining the new prognostic model parameters, it becomes possible to evaluate the degree of our knowledge, the magnitude of the chaotic factor, the complex probability, the *RUL* probability, and the system failure and survival probabilities; after that a pressure cycle time $t$ has been applied to the buried pipeline, which are all functions of the system degradation subject to chaotic and stochastic influences.

Accordingly, the advantages and the purpose of the current chapter are to:

1. Extend classical probability theory to the set of complex numbers and therefore to link the theory of probability to the field of complex variables and analysis. This job was started and elaborated in my previous 12 papers.

2. Do an updated follow-up of the degradation $D$ performance and behavior with cycle number or time which is subject to chaos. This follow-up is accomplished by the real failure probability of the system due to its definition that evaluates the jumps in $D$, therefore linking a system degradation to probability theory in a novel and original way.

3. Apply the new axioms of probability and paradigm to system prognostic; thus, I will extend the prognostic concepts to the set of complex probabilities $\mathcal{C}$.

4. Show that all stochastic phenomena can be expressed deterministically in the set of complex probabilities $\mathcal{C}$.

5. Measure and compute both the degree of our knowledge and the chaotic factor of the system remaining useful lifetime and its degradation.

6. Draw and illustrate the graphs of the parameters and functions of the original paradigm corresponding to a buried pipeline prognostic.

7. Show that the classical concepts of random remaining useful lifetime and degradation possess a probability permanently equal to one in the complex set; hence, no randomness, no chaos, no uncertainty, no ignorance, no disorder, and no unpredictability exist in:

$$\mathcal{C}\text{ (complex set)} = \mathcal{R}\text{ (real set)} + \mathcal{M}\text{ (imaginary set)}.$$

8. Show that by adding new and supplementary dimensions to any stochastic phenomenon, whether it is a pipeline system or any other random experiment, it becomes possible to do prognostic in a deterministic way in the set $\mathcal{C}$ of complex probabilities.

9. Pave the way to implement this novel model to other areas in stochastic processes and to the field of prognostics in science and engineering. These will be the topics of my future research works.

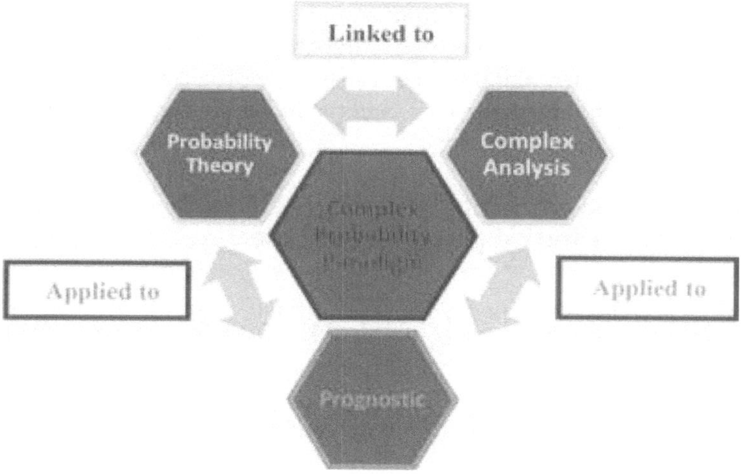


**Figure 1.**
*The diagram of the main purposes of the complex probability paradigm and research work.*

Concerning some applications of the original elaborated paradigm and as a future work, it can be applied to a wide set of dynamic systems like vehicle suspension systems and offshore and buried petrochemical pipelines which are subject to fatigue and in the cases of nonlinear and linear damage accumulation. Furthermore, compared with existing literature, the main contribution of the present research work is to apply the novel paradigm of complex probability to the concepts of random remaining useful lifetime and degradation of a buried pipeline system hence to the case of analytic prognostic in the case of linear damage accumulation subject to fatigue. The following figure shows the main purposes of the complex probability paradigm (*CPP*) (**Figure 1**).

To conclude and to summarize, in the real probability universe $\mathcal{R}$, our degree of our certain knowledge is regrettably imperfect; therefore we extend our study to the complex set $\mathcal{C}$ which embraces the contributions of both the real probabilities set $\mathcal{R}$ and the imaginary probabilities set $\mathcal{M}$. Subsequently, this will lead to a perfect and complete degree of knowledge in the universe $\mathcal{C} = \mathcal{R} + \mathcal{M}$ (since $Pc = 1$). In fact, working in the complex universe $\mathcal{C}$ leads to a certain prediction of any random event, because in $\mathcal{C}$ we eliminate and subtract from the calculated degree of our knowledge the quantified chaotic factor. This will yield a probability in the universe $\mathcal{C}$ equal to one ($Pc^2 = DOK - Chf = DOK + MChf = 1 = Pc$). Many

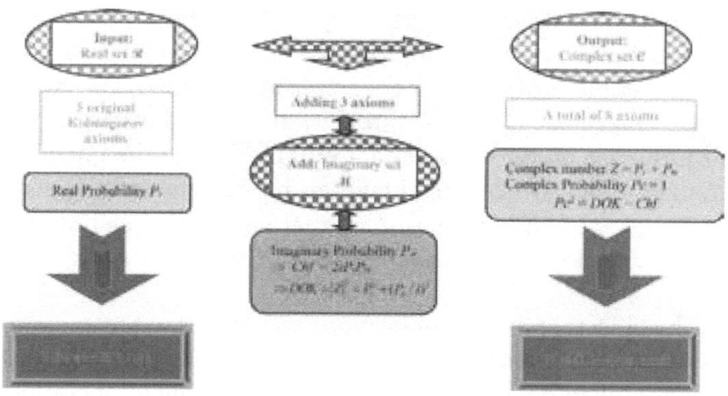

**Figure 2.**
*The EKA or the CPP diagram.*

illustrations considering various continuous and discrete probability distributions in my 12 previous research papers verify this hypothesis and novel paradigm [20–31]. The extended Kolmogorov axioms (*EKA* for short) or the complex probability paradigm can be summarized and shown in the following figure (**Figure 2**).

## 3. Previous research work: analytic prognostic and linear damage accumulation for buried petrochemical pipelines

In this section a comprehensive summary of a part of my previously published PhD thesis [16] and of the formerly published IFAC conference paper [14] will be done, and the results that this current chapter needs will be just cited.

### 3.1 A brief introduction to the adopted methodology

The objective of my earlier research study, which will be enhanced in the present chapter and will be linked to *CPP*, was to develop an analytic linear model of prognostic capable of predicting the remaining useful lifetime and the degradation *D* curves of a buried petrochemical pipeline system subject to fatigue starting from an initial known damage and under a given environment [14, 16]. This shows to be beneficial for many reasons which are fewer pipe bending; reduced plant congestion, wind, and other loads; and protection from ambient temperature changes. This work is restricted here to normal service loads that consist of only soil action and internal pressure.

Petrochemical pipelines are systems that are used to transport natural gas and oil between sites. We believe that pipeline tubes are a major element in petrochemical industries. As a matter of fact, the prognostic of their life is essential in this industry since their availability has decisive and critical consequences on the cost of exploitation. Fatigue, which is due to internal pressure-depression variation along time, is the major failure cause of these systems. These pipelines are typically devised for ultimate limit states (resistance). Additionally, due to soil aggression influences, buried pipelines are subject to corrosion. Pipelines are designed as cylindrical tubes of thickness *e* and radius *R*.

A target failure probability of about $10^{-5}$ for pipelines is suggested by the DNV 2000 rules. Their major failure causes are soil settlements, seismic ground waves, deformations, buckling, stress concentration in welding and fitting, internal and external corrosion, pressure fluctuation over a long period, and vibration and resonance. Moreover, crack detection tools detect the crack propagation caused by failures due to fatigue.

An important part of the main pipes is exposed to external cracking, which is a dangerous setback for the industry of pipes, for example, in the USA, Canada, and Russia. External crack identification is accomplished using diverse nondestructive evaluation (NDE) methods. If cracks were detected during inspection, we should evaluate their influence on the remaining useful lifetime of the pipeline in order to select the action of maintenance that should be applied: do nothing/repair/replace. We judge the integrity of pipes by assuming that some defects after in-line inspection (ILI) can be still undetected; detected, but not measured; detected and measured.

Moreover, the objective in my publications was to assess the evolution of the lifetime of a system at each instant. Consequently, and for this purpose, the trajectories of degradation had been utilized in terms of the time of operation or cycles' number. Hence, we deduce the *RUL* variations from these trajectories of degradation. Thus, I have considered many industrial illustrations in the simulation of my model in these earlier publications and work to prove the effectiveness of my model [1–6, 13–19]. Three case studies of pipelines were taken into consideration: buried, unburied, and subsea (offshore pipes). Each one of these situations necessitates different physical parameters like friction and soil pressure, corrosion, and atmospheric and water pressure. One of these cases is elaborated here which is the system of buried petrochemical pipes where three modes of pressure profiles (mode 1 = high, mode 2 = middle, and mode 3 = low-pressure conditions) were examined and simulated. My model showed that it presented a useful tool for a prognostic analysis and that it is very convenient in such industrial systems. Furthermore, it proved that it is less expensive than other models that require a huge number of measurements and data.

### 3.2 Fatigue crack growth

The stress intensity factor was introduced to calculate the correlation between the crack growth rate, $da/dN$, and the stress intensity factor range, $\Delta K$. The Paris-Erdogan's law [7] allows to evaluate the rate of propagation of the crack length $a$ after its detection. This damage growth law is expressed by the following equation:

$$\frac{da}{dN} = C(\Delta K)^m \tag{1}$$

where

$\frac{da}{dN}$, the crack growth rate = the increase of the crack length $a$ per cycle $N$.

$\Delta K(a) = Y(a)\Delta\sigma\sqrt{\pi a}$, the intensity factor of the stress.

$Y(a)$, the component's crack geometry function.

$\Delta\sigma$, the range of the applied stress in a cycle.

$m$ and $C$, the constants of materials obtained experimentally; $(2 \leq m \leq 4)$ and $(0 < C \ll 1)$.

### 3.3 The modeling of linear cumulative damage

To do the prognostic of a degrading element, my approach was to evaluate and to predict the end of life of the element by modeling and tracking the function of degradation. My model of damage, whose progress is up to the macro-crack initiation point, is illustrated in **Figure 3** by the damage linear rule of Palmgren-Miner.

As a matter of fact, this law [7] is used to calculate the cumulative damage $d_i$ of different stress levels $\sigma_i$ ($i$ = 1, $i$ = 2, ..., $i$ = $k$) applied for $n_i$ cycles. Knowing that $N_i$ is

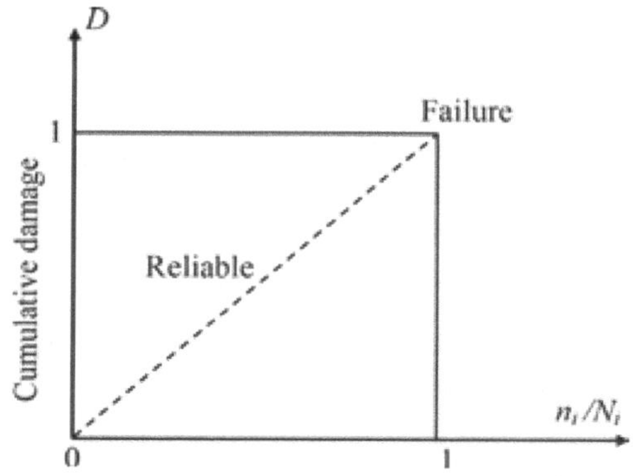


**Figure 3.**
*Palmgren-Miner's linear rule of damage.*

the total cycle's number of stress $\sigma_i$ to be applied and that lead to failure. The linear cumulative damage corresponding to the applied stresses ($i$ = 1 to $k$) is provided by

$$D_k = \sum_{i=1}^{k} d_i = \sum_{i=1}^{k} \frac{n_i}{N_i} \tag{2}$$

The initial detectable crack $a_0$ at the cycle $N_0$, the crack length $a_N$ at any cycle $N$, and the crack length $a_C$ at the failure cycle $N_C$ are estimated by a sensor, and their values are included in the model of damage prognostic in the equation of damage. It is expressed in my model by the resulting relation

$$D_N = \frac{a_N}{a_C - a_0} \tag{3}$$

Or in terms of the pressure cycle time $t$, the relation is given by

$$D_t = \frac{a_t}{a_C - a_0} \tag{4}$$

To simplify the study, it is suitable to adopt a measurement of damage denoted by $D \in [0, 1]$ which is computed by the Palmgren-Miner's law of linear cumulative damage. The damage level in a system at a specific cycle which is due to fatigue is illustrated by a scalar function of damage denoted by $D(t)$ or $D(N)$. "No damage" corresponds to the value $D$ = 0, and "total damage" or the appearance of the first macro-crack corresponds to $D$ = 1.

### 3.4 An expression for degradation

Therefore, my general prognostic analytic linear model function, which is a recursive relation for the sequence of $D$, is given by [16]

$$\begin{aligned} D_N &= D(N) = P_{rog}(a_N) \\ &= \frac{a_{N-1}}{a_C - a_0} + \frac{C}{a_C - a_0} \times (\pi a_{N-1})^{3/2} \times \left[0.6 \ \times \frac{1 + 2(a_{N-1}/e)}{(1 - a_{N-1}/e)^{\frac{3}{2}}}\right]^3 \times (P_j R/e)^3 \end{aligned} \tag{5}$$

where $C$, the environment parameter; $e$, the pipe thickness; $R$, the pipe radius; $a_0$, the initial crack length at the cycle $N_0$; $a_{N-1}$, the crack length at the load cycle N–1; $a_c$: the crack length at the failure cycle $N_C$. It was assumed in the model that $a_C = e/8$ for justified reasons [16]; $P_j$: the pipe internal pressure.

Or in terms of the pressure cycle time $t$, the recursive relation for the sequence of $D$ is given by

$$\begin{aligned} D_t &= D(t) = P_{rog}(a_t) \\ &= \frac{a_{t-1}}{a_C - a_0} + \frac{C}{a_C - a_0} \times (\pi a_{t-1})^{3/2} \times \left[0.6 \ \times \frac{1 + 2(a_{t-1}/e)}{(1 - a_{t-1}/e)^{\frac{3}{2}}}\right]^3 \times (P_j R/e)^3 \end{aligned} \tag{6}$$

Consequently, the previous recursive relation leads to a sequence of $D_t$ values with $N_0 \leq N \leq N_C$ or $t_0 \leq t \leq t_C$ whose limit is $D_C$ = 1:

$$D_0 = \frac{a_0}{a_C - a_0};\ D_1 = \frac{a_1}{a_C - a_0};\ D_2 = \frac{a_2}{a_C - a_0};\ \cdots;\ D_{t-1} = \frac{a_{t-1}}{a_C - a_0};\ \ D_t = \frac{a_t}{a_C - a_0} \tag{7}$$

We will consider three different levels of internal pressure to take into account the diverse states of pressure conditions which are low, middle, and high. Moreover, as the stress load is a function of the cycles $N$ or of time $t$, then we can draw the trajectories of degradation of $D(N)$ or $D(t)$ in addition to the trajectories of $RUL(N)$ or $RUL(t)$ in terms of the total number of loading cycles $N$ or in terms of the pressure time $t$. Therefore, my developed model of linear damage will be applied in order to compute the pipeline system prognostic.

## 3.5 The three levels of internal pressure simulations

We will consider in our current work a pipeline transporting natural gas of radius $R$ = 240 mm and of thickness $e$ = 8 mm. The parameters in this case are $C = 1.3 \times 10^{-14}$ (under soil, buried pipelines) and $m$ = 3 (metal). The initial crack length is considered to be $a_0$ = 0.02 mm. The crack length $a_C$ at the failure cycle time $t_C$ was assumed in the model to be equal to $e/8$ for justified reasons [16]. Hence, from Eqs. (5) and (6), we get

$$D_0 = \frac{a_0}{a_C - a_0} = \frac{a_0}{(e/8) - a_0} = \frac{0.02}{(8/8) - 0.02} = \frac{0.02}{0.98} = 0.020408163$$

The soil specific weight is $\gamma$ = 9.843 kN/m$^3$. The weight per linear meter of pipe and gas content is $W_p$ = 203.27 kg/m. The specific gravity of the pipe material and of the natural gas are, respectively, $\gamma_{pipe}$ = 7850 kg/m$^3$ and $\gamma_{gas}$ = 600 kg/m$^3$. The depth of the pipe is taken as $H = 7R$, and the friction coefficient interval is $0.5 \le \mu \le 0.7$ [16].

The internal pressure $P_j$ is modeled following a triangular form and distribution in order to be similar to the real case of pipeline operating condition (pressure-depression) (**Figure 4**).

We will consider three maximal levels of $P_j$ which are $P_0$ = 3, 5, and 8 MPa and with a period of repetition T. This repetition period varies depending on the conditions of exploitation; it is considered to be equal to 20 h. We note that these three levels are supposed to be the extreme conditions of the pipeline exploitations and are mean estimations of the real and actual random period and pressure rates. A trajectory of degradation $D(N)$ is inferred at each of these three levels in terms of the cycle number $N$ or pressure cycle time $t$. When $D_t$ or $D_N$ attains the unit value, therefore the corresponding $t = t_C$ or $N = N_C$ is the lifetime of the pipeline in the fatigue case.

For the purposes of simulations, in **Table 1**, the values of pressure $P_j$ are considered to be equal to the maximal values $P_0$. The analytic linear prognostic model

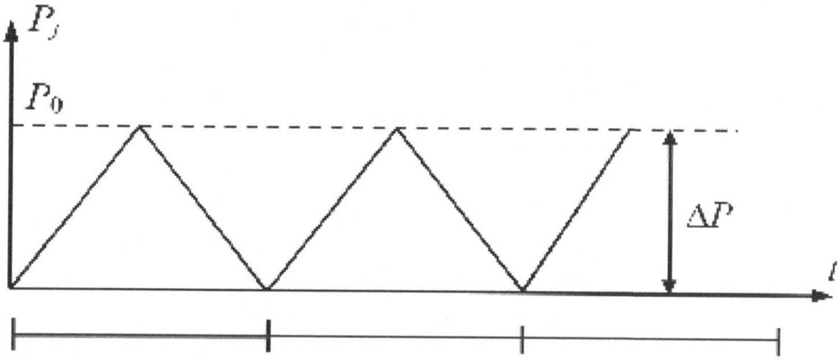


**Figure 4.**
*Triangular variation of internal pressure.*

| Pressure mode | $P_j$ (MPa) | Model |
|---|---|---|
| High (mode 1) | 8 | Triangular |
| Middle (mode 2) | 5 | Triangular |
| Low (mode 3) | 3 | Triangular |

**Table 1.**
*Characteristics of each internal pressure mode.*

(Eqs. (5) and (6)) simulation is achieved for each internal pressure level (low, middle, and high).

A huge amount of pressure simulations of the order of hundreds of millions are required to estimate the real system lifetime; hence, we have used an approximated model of lifetime simulation of the order of 10,000,000 iterations. Accordingly, we have considered for this purpose a high-capacity computer system: a workstation computer with parallel microprocessors, a 64-Bit operating system, a 64 GB RAM, as well as a 64-Bit MATLAB version 2019 software.

### 3.6 RUL computation

The evaluation of the remaining useful lifetime of the system is the major objective in a prognostic study. Since the *RUL* is the complement of the damage curve $D(t)$, it can be deduced from it. Accordingly, at each time $t$, the required *RUL* is the length from cycle time $t$ to the critical cycle time $t_C$ that corresponds to the threshold $D = 1$. The entire *RUL* is inferred using the following relation:

$$RUL = t_C - t_0 \tag{8}$$

where $t_C$ is the necessary cycle time for the appearance of the first macro-cracks that means to reach failure, and $t_0$ is the initial cycle time considered in general to be equal to 0.

Consequently, my prognostic model computes the *RUL*s for the three internal pressure modes that can be now simply inferred from these three curves at any instant $t$ or at any active cycle $N$ in this manner:

For mode 3, $RUL_3(t) = t_{C3} - t$.
For mode 2, $RUL_2(t) = t_{C2} - t$.
For mode 1, $RUL_1(t) = t_{C1} - t$.

### 3.7 The effects of environment in the suggested prognostic model

Two parameters which are $C$ and $m$ embody the effects of the environment. These two parameters are associated to the material environment. $C$ and $m$ depend on the initial crack length, on the geometry and size of the specimen, and on the testing conditions (such as the loading ratio $\sigma$). These two parameters affect the performance of the material during the process of fatigue through the crack propagation. The influencing parameters on this fatigue process, like humidity, temperature, material nature, geometry dimensions, applied load location, corrosion, water action, soil pressure and friction, atmospheric pressure, etc., can be stochastic and can be also embodied by $C$ and $m$. Furthermore, it is crucial to note here that these two parameters can be as well random variables and hence can be represented by probability distributions materializing the environment stochastic and chaotic influences on the system. It is also important to mention that these two parameters are computed by the mean of experiments in real conditions. We give here some examples from several and other prognostic studies [7, 8]: $C = 5.2 \times 10^{-13}$ (free air, unburied pipelines), $C = 1.3 \times 10^{-14}$ (under soil, buried pipelines), $C = 2 \times 10^{-11}$ (for offshore pipelines), and $m = 3$ (metal).

## 4. The complex probability paradigm applied to prognostic

In this section, the novel complex probability paradigm will be presented after applying it to prognostic.

### 4.1 The basic parameters of the new model

It is very well-known that in systems engineering, the remaining useful lifetime and the degradation prediction is profoundly linked to many aspects (like humidity, temperature, material nature, geometry dimensions, applied load location, water action, corrosion, soil pressure and friction, atmospheric pressure, etc.) that usually have a stochastic and chaotic behavior which reduces the degree of our certain system knowledge [32–35]. Consequently, the lifetime of the system becomes a random variable and is computed by the arbitrary time $t_C$ which is evaluated when sudden failure occurs due to these stochastic causes and chaotic factors. We can deduce from the *CPP* that we can foretell the exact probabilities of *RUL* and *D* with certitude in the whole set $\mathcal{C} = \mathcal{R} + \mathcal{M}$ if we add to the probability measure of a random variable in the real set $\mathcal{R}$ the corresponding imaginary counterpart $\mathcal{M}$ since $Pc = 1$ perpetually and constantly. In fact, prognostic is based on the forecast of a system remaining useful lifetime at any cycle $N$ or instant $t$ and during the system operation. Therefore, we can make use of this novel idea and procedure to do the prognostic analysis of the system *RUL* and degradation prediction and evolution.

Let us consider a system degradation trajectory $D(t)$ where we study a specific instant (or cycle) $t_k$. The system age is measured by the number of years and by the variable $t_k$ (**Figure 5**). From the illustrated figures (**Figures 5a** and **5b**), we can infer that at the system age $t_k$ of the prognostic study must give the prediction of the failure instant $t_C$. Therefore, the *RUL* predicted here at the instant $t_k$ has the following value:

$$RUL(t_k) = t_C - t_k \quad (9)$$

As a matter of fact, at $t_k = 0$ (at the beginning) (point J), the system is intact, then the failure probability of the system is $P_r = 0$, the chaotic factor in our prognostic is null ($MChf = 0$) because no chaos exists yet, and our knowledge of the unharmed and undamaged system is complete and certain ($DOK = 1$); consequently,

$$RUL(0) = t_C - t_k = t_C - 0 = t_C.$$

If $t_k = t_C$ (point L), the system is completely damaged, then $RUL(t_C) = t_C - t_C = 0$, and therefore the failure probability of the system is one ($P_r = 1$). Failure occurs at this point. Thus, our knowledge of the totally worn-out system is perfect ($DOK = 1$) and the harmful task of chaos has finished; hence it is no more applicable ($MChf = 0$).

If $0 < t_k < t_C$ (point K, where J < K < L), the probability of occurrence of this instant and the probabilities of prediction of *RUL* and *D* are both less than 1 and are imperfect in $\mathcal{R}$ ($0 < P_r < 1$). This is the result of non-zero chaotic factors influencing the system ($MChf > 0$). The system degree of our knowledge which is subject to chaos is thus uncertain and is consequently less than one in $\mathcal{R}$ ($0.5 < DOK < 1$).

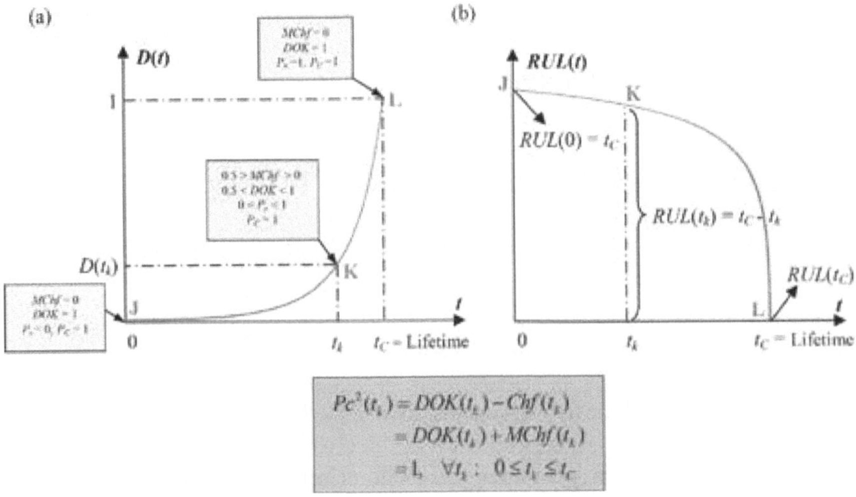


**Figure 5.**
*CPP. (a) The prognostic of degradation and (b) The prognostic of RUL.*

Furthermore, by applying here the *CPP* paradigm, we can therefore determine at any instant $t_k$ $(0 \leq t_k \leq t_C)$ and, at any point between J and L inclusively, the *RUL* and $D$ of the system with certitude in the set $\mathcal{C} = \mathcal{R} + \mathcal{M}$ because in $\mathcal{C}$ we have $Pc = 1$ permanently.

Additionally, we can express two complementary phenomena or events $E$ and $\overline{E}$ by their respective probabilities as follows:

$$P_{rob}(E) = p \text{ and } P_{rob}\left(\overline{E}\right) = q = 1 - p.$$

Therefore, let the probability $P_{rob}(E)$ as a function of the time $t_k$ be defined by

$$P_{rob}(E) = P_{rob}(t \leq t_k) = F(t_k) \tag{10}$$

where the classical and usual cumulative distribution function (*CDF*) of the random variable $t$ is denoted by the term $F(t)$.

Since $P_{rob}(E) + P_{rob}\left(\overline{E}\right) = 1$, therefore, we deduce at an instant $t = t_k$:

$$P_{rob}\left(\overline{E}\right) = 1 - P_{rob}(E) = 1 - P_{rob}(t \leq t_k) = P_{rob}(t > t_k) = 1 - F(t_k) \tag{11}$$

In addition, two particular instants can be defined:

$t = t_0 = 0$ which corresponds to the system raw state and which is assumed to be the initial time of functioning where $D = D_0$.

$t = t_C$ which corresponds to the system wear-out state and which is the failure instant where $D = D_C = 1$.

Consequently, we can state the boundary conditions as follows:

For $t = t_0 = 0$, we have $D = D_0 \approx 0$ (the initial damage that may be nearly 0) and $F(t) = F(t_0) = P_{rob}(t \leq 0) = 0$.

For $t = t_C$, we have $D = D_C = 1$ and $F(t) = F(t_C) = P_{rob}(t \leq t_C) = 1$.

We note also that since $F(t_k)$ is defined as a cumulative probability function, then $F(t_k)$ is a non-decreasing function that varies between 0 and 1. In addition, since $RUL(t_k) = t_C - t_k$ and $t_k$ is always increasing $(0 \leq t_k \leq t_C)$, then $RUL(t_k)$ is a non-increasing remaining useful lifetime function (**Figure 5b**).

## 4.2 The new prognostic model

The novel model of prognostic basic assumption will be presented now [36–53]. We assume first the cumulative probability distribution function $F(t)$ of the random variable time $t$ as being equal to the function of degradation itself, which means

$$F(t_k) = P_{rob}(t_0 \le t \le t_k) = \sum_{t=t_0}^{t=t_k} P_{rob}(t) = D(t_k) \tag{12}$$

We mention here that we are working with discrete random functions that depend on the discrete random time $t$ of pressure cycles.

This basic assumption is reasonable because:

1. Both $D$ and $F$ are cumulative functions starting from zero and ending with one.

2. Both are non-decreasing functions.

3. Both functions are without measure units: $D$ is an indicator quantifying system damage and degradation, as well as $F$ which is an indicator quantifying randomness and chance.

Afterward, we suppose that, at the instant $t = t_k$, the term $P_r(t)/\psi_j$ is the real probability of system failure and is computed as follows:

$$\begin{aligned} P_r(t_k) &= \psi_j \times [P_{rob}(t \le t_k) - P_{rob}(t \le t_{k-1})] = \psi_j \times [F(t_k) - F(t_{k-1})] \\ &= \psi_j \times [D(t_k) - D(t_{k-1})] \\ &= \psi_j \times \left[\sum_{t=t_0}^{t=t_k} P_{rob}(t) \;-\; \sum_{t=t_0}^{t=t_{k-1}} P_{rob}(t)\right] \\ &= \psi_j \times \sum_{t=t_{k-1}}^{t=t_k} P_{rob}(t) = \psi_j \times P_{rob}(t_{k-1} \le t \le t_k) \end{aligned} \tag{13}$$

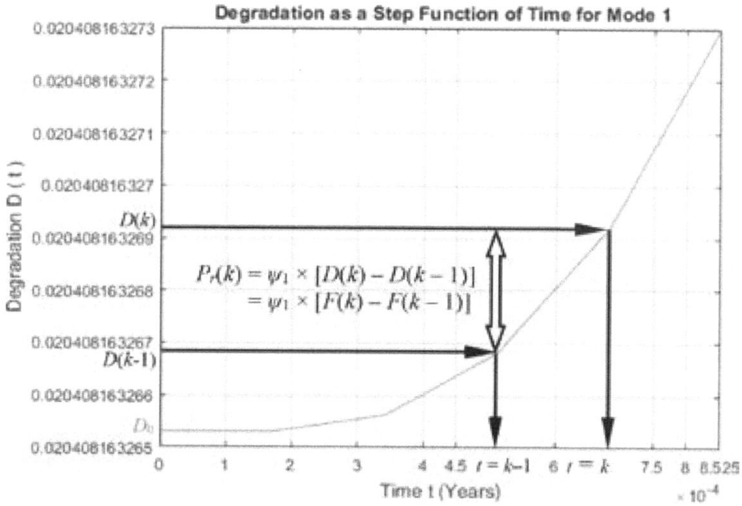


**Figure 6.**
*$P_r$, degradation, and the* CDF *step function.*

= $\psi_j$ times the jump in $F(t)$ or $D(t)$ from $t = t_{k-1}$ to $t = t_k$ (**Figures 6** and 7).

where $t = [0, 1, 2, \dots, t_{k-1}, t_k, t_{k+1}, \dots, t_C]$ is the time of pressure cycles and $t_0 = 0$ is the initial time of pressure cycles at the simulation beginning. It corresponds to a degradation $D = D(t_0) = D_0$ which is generally considered to be nearly equal to 0.

Hence, since $F(t_k) = D(t_k)$ then $F(t_0) = D(t_0) = 0.020408 \approx 0$, but $F(t_0)$ is taken all over this research work as being equal to 0;

$t_1 = 1$ = the first pressure cycle time ... $t_k$ = the $k^{\text{th}}$ pressure cycle time ... $t_C$ = the pressure cycles time that leads to system failure = the critical pressure time. It corresponds to $D = D_C = 1$. It follows directly that $F(t_C) = D(t_C) = D_C = 1$.

$\psi_j$ is the simulation magnifying factor that depends on the pressure profile. It is $\psi_1 = 5082$ for the high-pressure mode ($j$ = 1, mode 1), $\psi_2 = 6737$ for the middle-pressure mode ($j$ = 2, mode 2), and $\psi_3 = 9151$ for the low-pressure mode ($j$ = 3, mode 3).

Thus, initially we have

$$P_r(t_k = t_0 = 0) = \psi_j \times F\,(t_0) = \psi_j \times 0 = 0$$

Moreover,

$$P_r(t_k) = \psi_j \times f_j(t_k) \Rightarrow P_r(t_k)/\psi_j = f_j(t_k), \tag{14}$$

where $1/\psi_j$ is a normalizing constant that is used to reduce $P_r(t_k)$ function to a probability density function (PDF) with a total probability equal to one. $1/\psi_j$ is a function of the pressure mode and conditions, and it depends on the parameters in the degradation (Eqs. (5) and (6)). The decreasing values of $1/\psi_j$ are logical since pipeline failure probabilities are decreasing with the decreasing pressure modes; hence, $1/\psi_1 > 1/\psi_2 > 1/\psi_3$. Consequently, we deduce that $f_j(t_k)$ is the usual probability density function (*PDF*) for each pressure mode $j$. Knowing that, from classical probability theory, we have always:

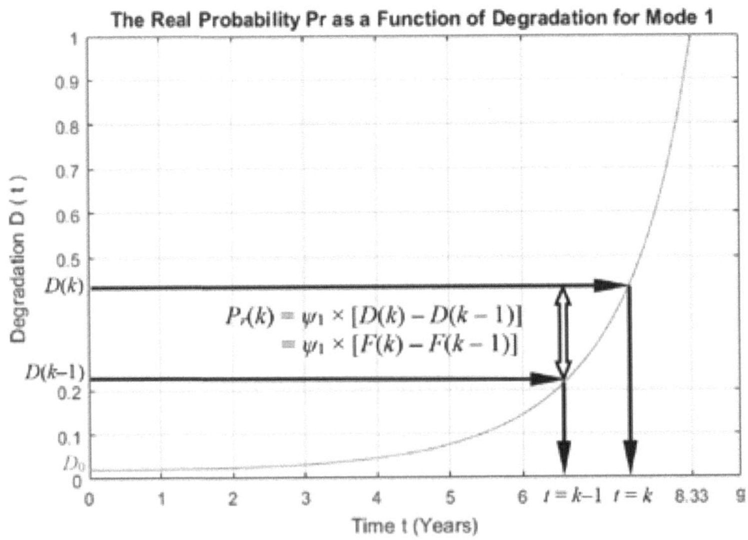


**Figure 7.**
*$P_r$ as a function of degradation $D(t)$.*

$$\sum_{t_k=t_0}^{t_k=t_C} f_j(t_k) = \sum_{t_k=t_0}^{t_k=t_C} P_r(t_k)/\psi_j = 1 \text{ for any pressure profile } j = 1, 2, 3.$$

This result is reasonable since $P_r(t_k)/\psi_j$ is here a probability density function (**Figure 6**).

Therefore, we can deduce that

$$\begin{aligned}\sum_{t_k=t_0}^{t_k=t_C} P_r(t_k) &= \psi_j \times \sum_{t=t_0}^{t=t_C} P_{rob}(t) = \psi_j \times P_{rob}(t_0 \leq t \leq t_C)\\ &= \psi_j \times [F(t=t_C) - F(t=t_0)] = \psi_j \times [D(t=t_C) - D(t=t_0)]\\ &= \psi_j \times F(t_C) \approx \psi_j \times D(t_C),\end{aligned}$$

$$\text{since } D(t_C) = 1 \text{ and } D(t_0) = 0.020408 \approx 0 \text{ and } F(t_0) \text{ is taken as } = 0 \tag{15}$$

$$= \psi_j \times \sum_{t_k=t_0}^{t_k=t_C} f_j(t_k) = \psi_j \times 1 = \psi_j$$

$$\Rightarrow \sum_{t_k=t_0}^{t_k=t_C} P_r(t_k)/\psi_j = 1, \text{ for any pressure profile } j = 1, 2, 3$$

We can understand that $F(t) = D(t)$ is a discrete *CDF* where the amount of the jump is $P_r(t)/\psi_j$; then, $P_r(t)/\psi_j$ is a damage evolution and degradation function (**Figures 6** and 7). And we can infer from the preceding computations that $P_r(t)/\psi_j$ is a probability density function. Accordingly, we can realize now that $P_r(t)/\psi_j$ quantifies and measures the system degradation or failure probability. Consequently, what we have achieved at this point is that we have linked degradation measure to probability theory.

We can notice the following:

$$0 \leq P_r(t_k)/\psi_j \leq 1, 0 \leq F(t_k) \leq 1, \text{and } (D_0 \approx 0) \leq D(t_k) \leq (D_C = 1),$$
$$\text{for every } t_k : 0 \leq t_k \leq t_C.$$

and

If $t_k \to 0 \Rightarrow D \to D_0 = 0.020408 \approx 0 \Rightarrow F \to 0 \Rightarrow P_r(t_k) \to 0$

if $t_k \to t_C \Rightarrow D \to D_C = 1 \Rightarrow F \to 1 \Rightarrow P_r(t_k) \to 1$.

This, since the degradation is very flat near 0 and starts increasing with $t$, becoming very acute at $t = t_C$, hence, near $t_C$, $P_r$ is the greatest and is equal to 1 (**Figures** 7 and 8).

Furthermore, we have:

$RUL(t_k) = t_C - t_k$ and it corresponds to a degradation of $D(t_k)$.

$RUL(t_{k-1}) = t_C - t_{k-1}$ and it corresponds to a degradation of $D(t_{k-1})$.

This implies that (**Figure 9**)

$$\begin{aligned}P_r(t_k) &= \psi_j \times [D(t_k) - D(t_{k-1})]\\ &= \psi_j \times \{D[t_C - RUL(t_k)] - D[t_C - RUL(t_{k-1})]\}\end{aligned} \tag{16}$$

### 4.3 Analysis and extreme chaotic and random conditions

Although the analytic linear laws of prognostic are very well-known and deterministic in [14, 16], there are general influences and aspects that can be chaotic and stochastic (like humidity, temperature, material nature, geometry dimensions, applied load location, water action, corrosion, soil pressure and friction, atmospheric pressure, etc.). Moreover, various variables in the expressions (5) and (6) of degradation which are considered as deterministic can also have a random aspect, such as the magnitude of applied pressure (due to the different conditions of pressure profile) and the length of the initial crack (potentially existing from the process of manufacturing). All those stochastic factors, embodied in the model by their mean values, influence the buried pipeline system and make its function of

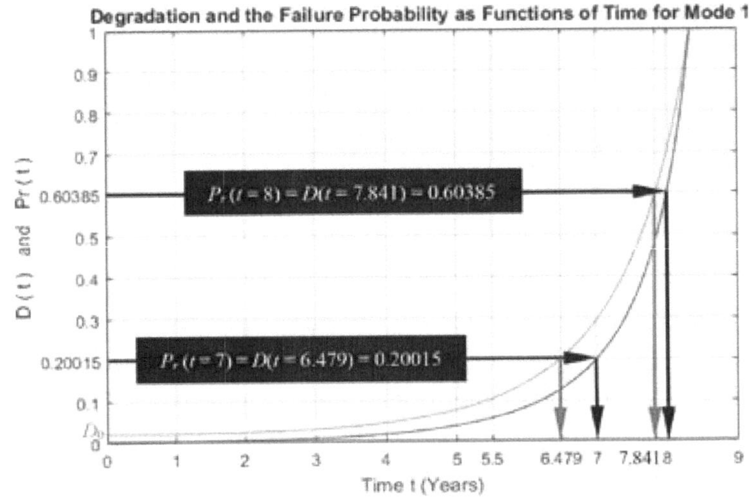


**Figure 8.**
*Degradation and $P_r$.*

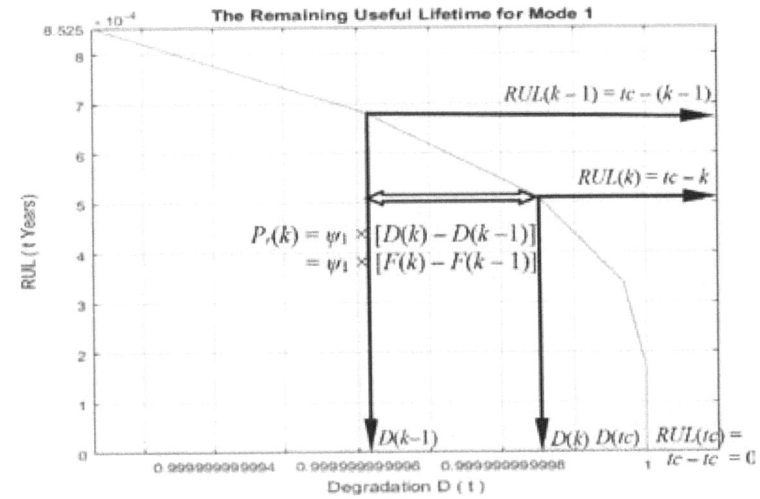


**Figure 9.**
*$P_r$, D, and RUL.*

degradation diverge from its computed trajectory modeled by these deterministic laws. An updated follow-up of the degradation performance and behavior with cycle number or time, which is subject to non-chaotic and chaotic influences, is made possible by $P_r(t_k)/\psi_j$ due to its definition that evaluates the jumps in $D$. In fact, chaos modifies and affects all the environment and system parameters included in the degradation equations (Eqs. (5) and (6)). Consequently, chaos total effect on the pipelines contributes to shape the degradation curve $D$ and is materialized by and counted in the pipeline system failure probability $P_r(t_k)/\psi_j$. Actually, $P_r(t_k)/\psi_j$ quantifies the resultant of all the nonrandom (deterministic) and random (nondeterministic) parameters and aspects which are contained in the equation of $D$, which affect the system and which lead to the consequent final curve of degradation. Consequently, an accentuated influence of chaos on the pipeline can lead to a smaller (or bigger) jump in the trajectory of degradation and therefore to a smaller (or bigger) failure probability $P_r(t_k)/\psi_j$. If, for example, due to extreme deterministic causes and random factors, $D$ jumps directly from $D_0 \approx 0$ to 1 then $RUL$ goes straight from $t_C$ to 0 and consequently $P_r(t_k)/\psi_j$ jumps instantly from 0 to 1:

$$P_r(t_k)/\psi_j = D(t_k) - D(t_{k-1}) = D(t_C) - D(0) \approx 1 - 0 = \sum_{t=0}^{t=t_C} P_{rob}(t) = 1$$

where $t$ jumps directly from 0 to $t_C$.

In the extreme ideal case, if the pipeline system never deteriorates (no stresses or pressure) and with zero random causes and chaotic factors, then the resultant of all the nondeterministic and deterministic influences is null (like in the pipeline isolated and idle state). Accordingly, the system remains indefinitely at $D_0 \approx 0$ and $RUL$ stays equal to $t_C$. So consequently, the jump in $D$ is constantly zero. Hence, the failure probability remains ideally 0:

$$P_r(t_k)/\psi_j = [D(t_k) - D(t_{k-1})] = [D_0 - D_0] = 0$$

where $D(t_0) = D(t_1) = \ldots = D(t_{k-1}) = D(t_k) = D(t_{k+1}) = \ldots = D_0 = 0.020408 \approx 0$, for $k = 0, 1, 2, 3, \ldots \infty$.

**Figure 6** illustrates the real probability of failure $P_r(t)$ in terms of the random degradation step $CDF$ of the pipeline as a function of the cycle time $t$ of pressure for mode 1.

**Figure 7** illustrates the real probability of failure $P_r(t)$ in terms of the random degradation of the pipeline as a function of the cycle time $t$ of pressure for mode 1.

**Figure 8** illustrates the real probability of failure $P_r(t)$ and the random degradation $D(t)$ of the pipeline in terms of the number of cycle time $t$ of pressure for mode 1.

**Figure 9** illustrates the real probability of failure $P_r(t)$ in terms of the random degradation $D(t)$ of the pipeline and the random $RUL(t)$ of the pipeline as a function of the cycle time $t$ (in years) of pressure for mode 1.

### 4.4 The flowchart of the complex probability analytic linear prognostic model

The following flowchart summarizes all the procedures of the proposed complex probability prognostic model:

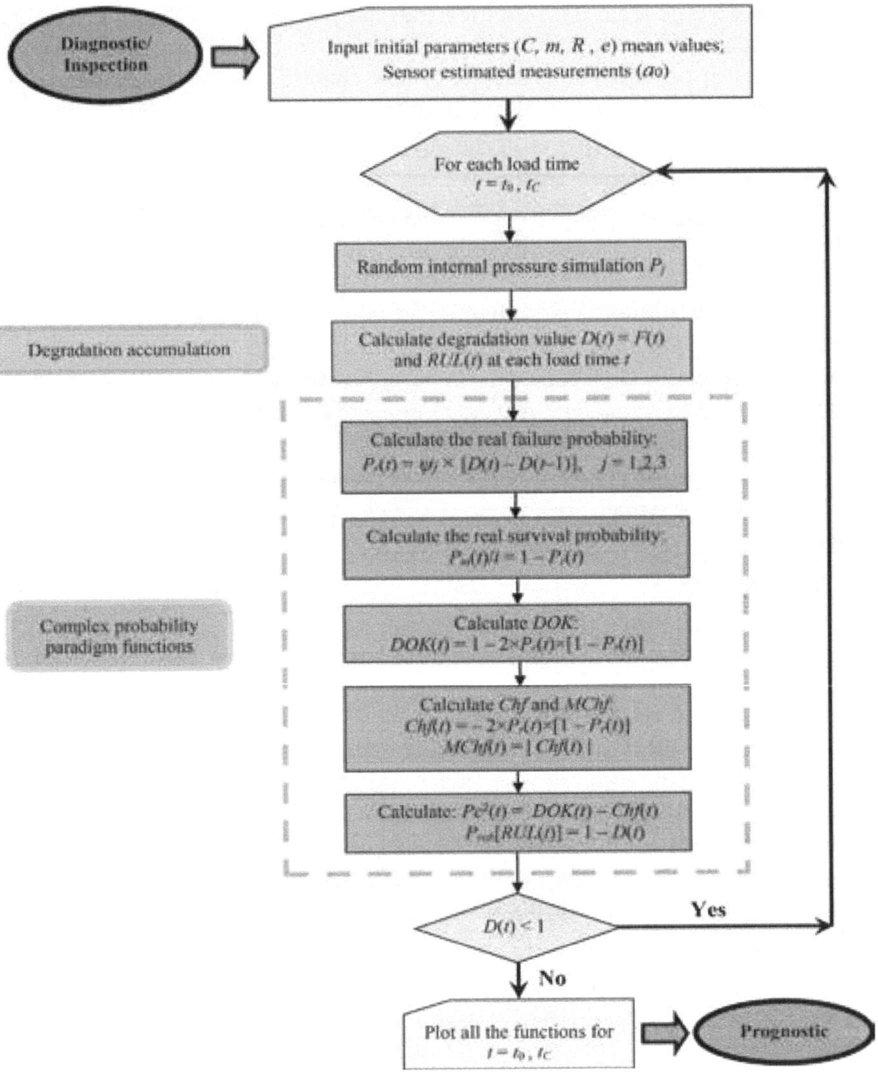


## 4.5 The evaluation of the new paradigm parameters

We can infer from what has been elaborated previously the following:

The real probability is $P_r(t_k) = \psi_j \times [D(t_k)–D(t_{k-1})]$, for pressure modes $j = 1, 2, 3$ (17)

The imaginary probability is $P_m(t_k) = i \times [1–P_r(t_k)] = i \times \left\{1–\psi_j \times [D(t_k)–D(t_{k-1})]\right\}$ (18)

The complementary probability is $P_m(t_k)/i = 1–P_r(t_k) = 1–\psi_j \times [D(t_k)–D(t_{k-1})]$ (19)

The complex probability vector is $Z(t_k) = P_r(t_k) + P_m(t_k) = P_r(t_k) + i \times [1–P_r(t_k)]$ (20)

The degree of our knowledge

$$DOK(t_k) = |Z(t_k)|^2 = 1 + 2iP_r(t_k)P_m(t_k) = 1\text{–}2P_r(t_k)P_m(t_k)/i = 1\text{–}2P_r(t_k)[1\text{–}P_r(t_k)] \\ = 1\text{–}2P_r(t_k) + 2P_r^2(t_k) \tag{21}$$

The chaotic factor

$$Chf(t_k) = 2iP_r(t_k)P_m(t_k) = -2P_r(t_k)P_m(t_k)/i = -2P_r(t_k)[1 - P_r(t_k)] \\ = -2P_r(t_k) + 2P_r^2(t_k) \tag{22}$$

*Chf* is null when $P_r(N_k) = P_r(0) = 0$ (point J) and when $P_r(t_k) = P_r(t_C) = 1$ (point L) (**Figures 5a** and **5b**).

The magnitude of the chaotic factor *MChf*:

$$MChf(t_k) = |Chf(t_k)| = -2iP_r(t_k)P_m(t_k) = 2P_r(t_k)P_m(t_k)/i = 2P_r(t_k)[1 - P_r(t_k)] \\ = 2P_r(t_k) - 2P_r^2(t_k) \tag{23}$$

*MChf* is null when $P_r(t_k) = P_r(0) = 0$ (point J) and when $P_r(t_k) = P_r(t_C) = 1$ (point L) (**Figures 5a** and **5b**).

At any instant $t_k$ $0 \leq \forall t_k \leq t_C$, the probability expressed in the complex set $\boldsymbol{C}$ is the following:

$$\begin{aligned} Pc(t_k)^2 &= [P_r(t_k) + P_m(t_k)/i]^2 = |Z(t_k)|^2 - 2iP_r(t_k)P_m(t_k) \\ &= DOK(t_k) - Chf(t_k) \\ &= DOK(t_k) + MChf(t_k) \\ &= 1 \end{aligned} \tag{24}$$

then, $Pc(t_k) = P_r(t_k) + P_m(t_k)/i = P_r(t_k) + [1 - P_r(t_k)] = 1$ always.

Therefore, the prognostic of $RUL(t_k)$ and $D(t_k)$ of the pipeline in the set $\boldsymbol{C}$ is forever certain. The buried pipeline system is considered thereafter under three modes of pressure in order to simulate the cumulative distribution function $D(t_k) = F(t_k)$ and hence in order to visualize, to quantify, as well as to draw all the prognostic parameters and *CPP*.

## 5. The simulation of the new paradigm

We will simulate in this section the original model of prognostic for the three internal pressure modes. We note that we have used the 64-Bit MATLAB version 2019 software to evaluate and find all the numerical values of the paradigm functions analysis.

### 5.1 The parameter simulation in the pipeline prognostic for mode 1

See **Figures 10–12**.

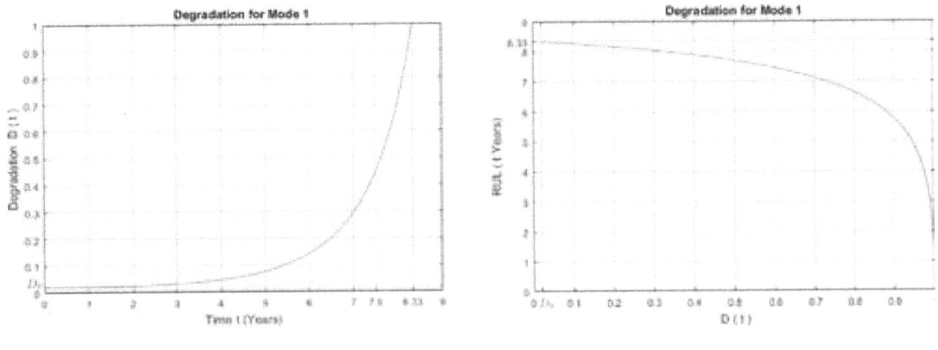


**Figure 10.**
*Pipeline degradation (a) and* RUL *(b) under linear damage law for high-pressure mode of excitation (mode 1).*

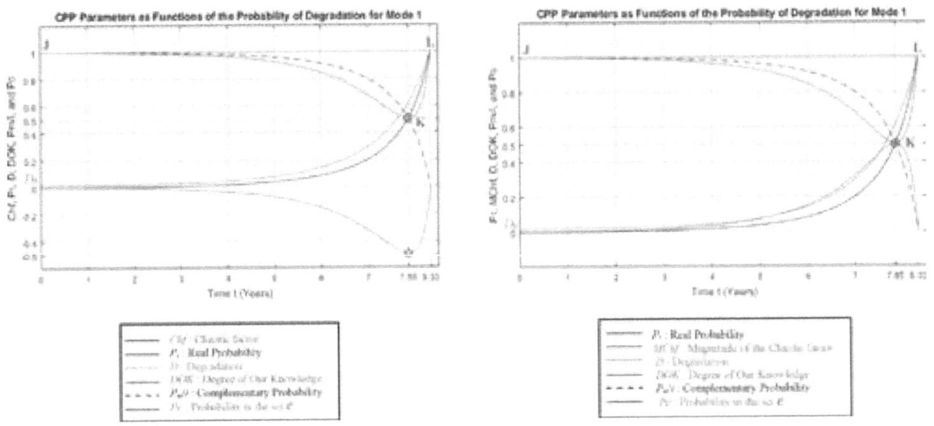


**Figure 11.**
*Degradation and* CPP *parameters with* Chf *(a) and with* MChf *(b) for mode 1.*

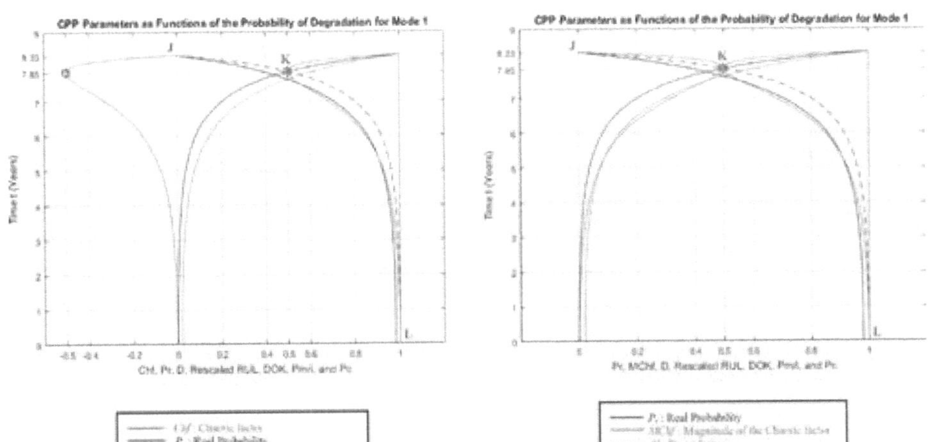


**Figure 12.**
*Degradation, rescaled* RUL, *and* CPP *parameters with* Chf *(a) and with* MChf *(b) for mode 1.*

### 5.1.1 The complex probability cubes for mode 1

See **Figures 13–15.**

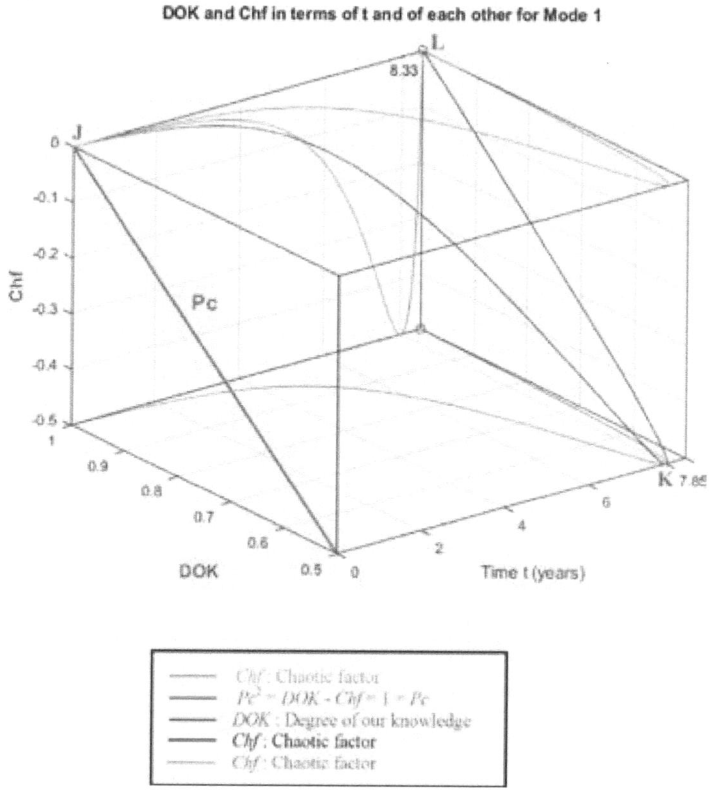


**Figure 13.**
*DOK and Chf in terms of t and of each other for mode 1.*

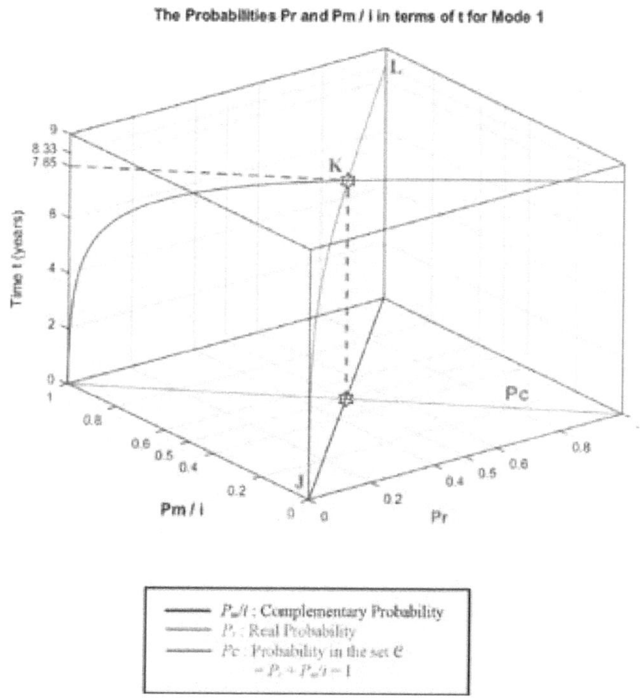


**Figure 14.**
*$P_r$ and $P_m/i$ in terms of t and of each other for mode 1.*

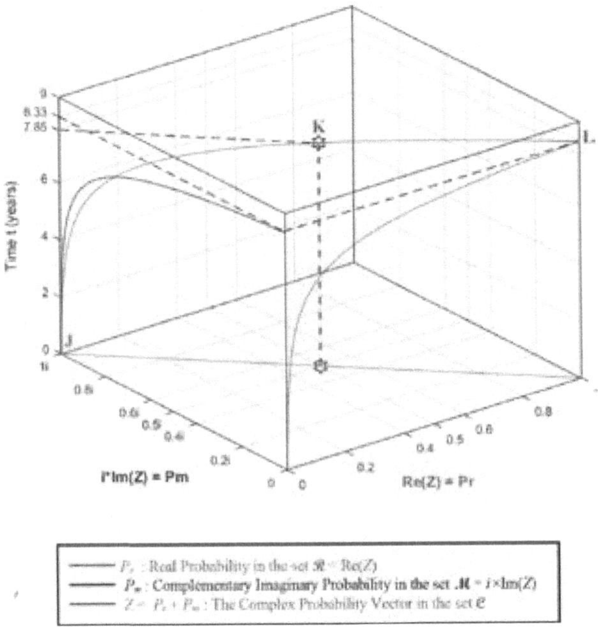


**Figure 15.**
*The complex probability vector Z in terms of t for mode 1.*

## 5.2 The parameter simulation in the pipeline prognostic for mode 2

See **Figures 16–18.**

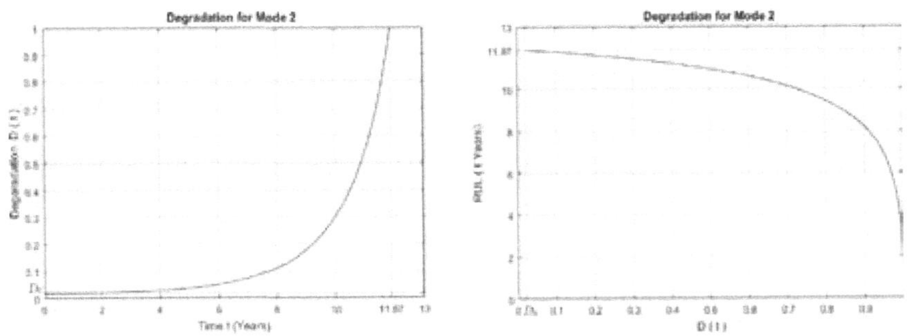


**Figure 16.**
*Pipeline degradation (a) and* RUL *(b) under linear damage law for middle-pressure mode of excitation (mode 2).*

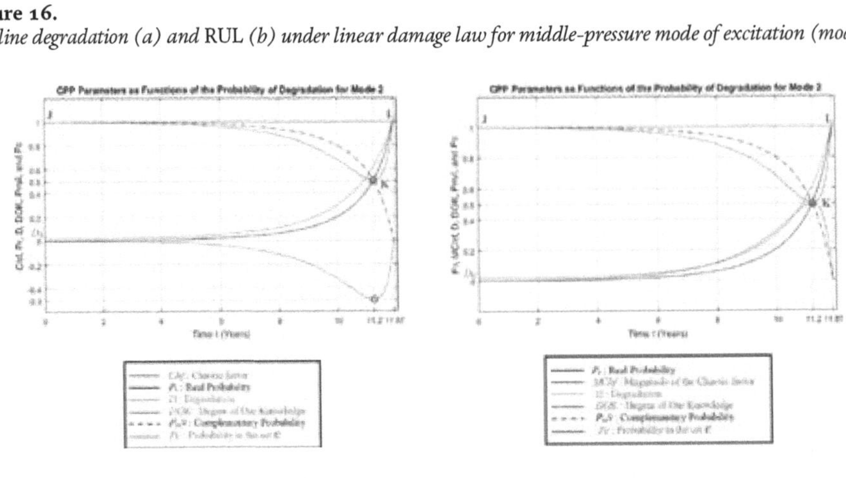


**Figure 17.**
*Degradation and* CPP *parameters with* Chf *(a) and with* MChf *(b) for mode 2.*

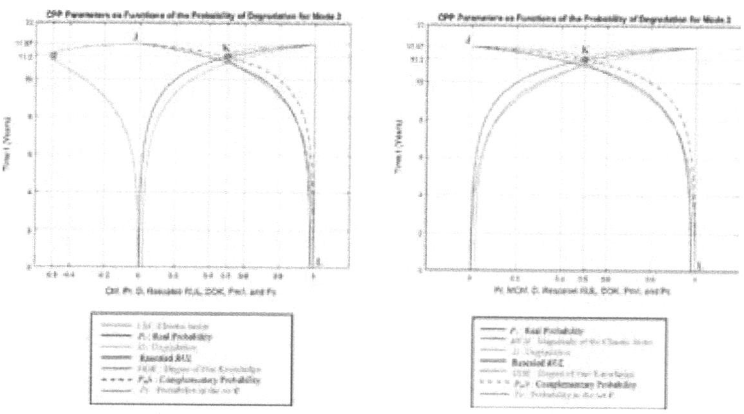

**Figure 18.**
*Degradation, rescaled* RUL, *and* CPP *parameters with* Chf *(a) and with* MChf *(b) for mode 2.*

### *5.2.1 The complex probability cubes for mode 2*

See **Figures 19–21.**

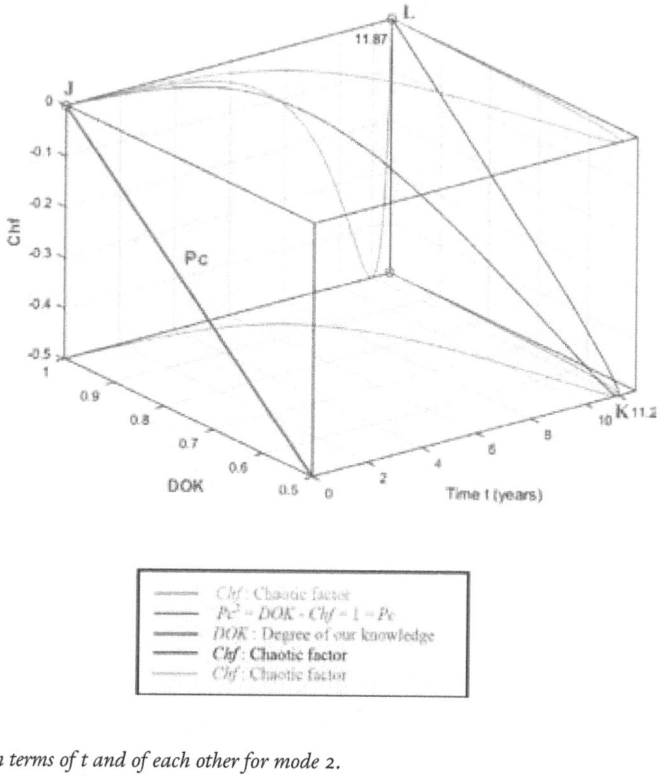


**Figure 19.**
*DOK and Chf in terms of t and of each other for mode 2.*

## 5.3 The parameter simulation in the pipeline prognostic for mode 3

See **Figures 22–24.**

### *5.3.1 The complex probability cubes for mode 3*

See **Figures 25–27.**

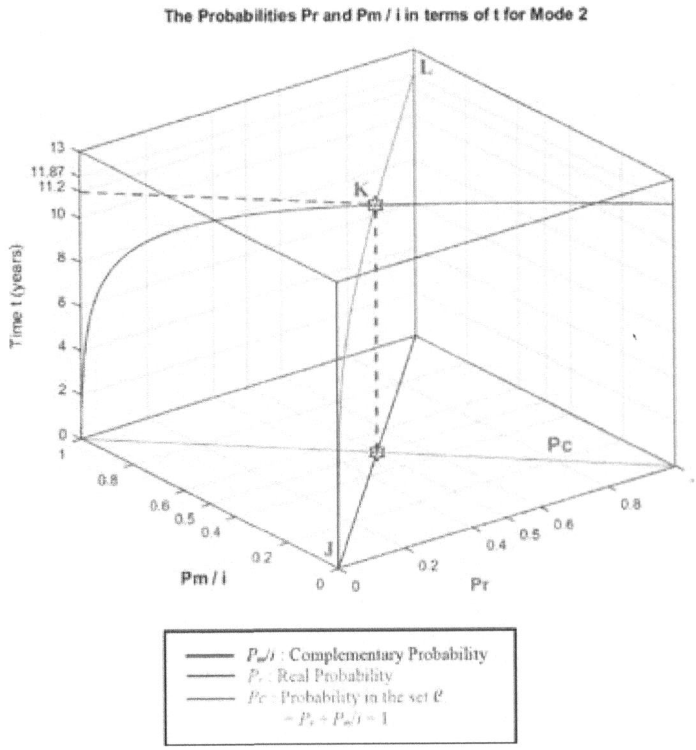


**Figure 20.**
*$P_r$ and $P_m/i$ in terms of t and of each other for mode 2.*

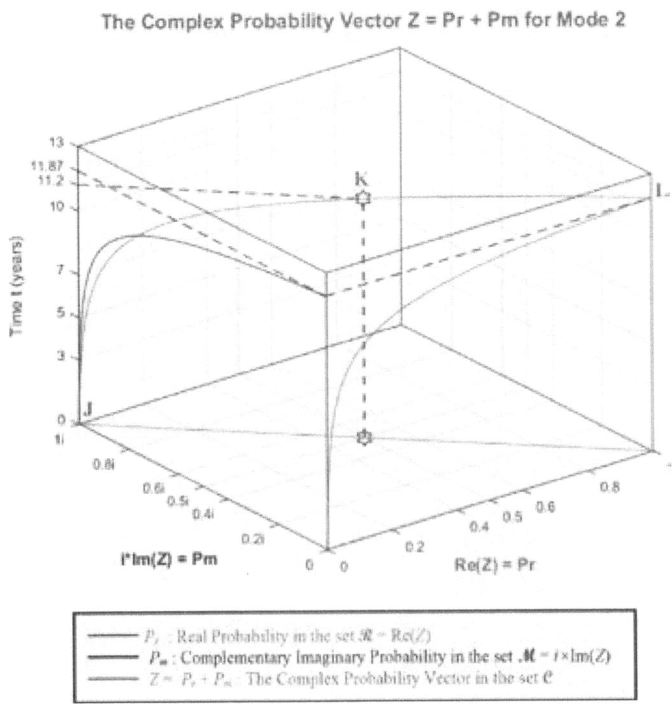


**Figure 21.**
*The complex probability vector Z in terms of t for mode 2.*

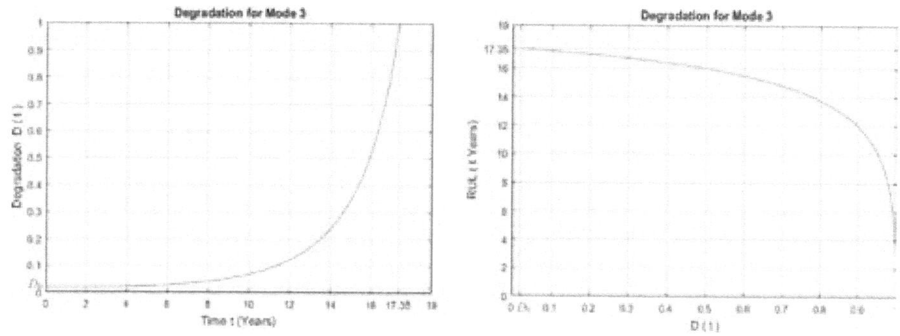


**Figure 22.**
*Pipeline degradation (a) and RUL (b) under linear damage law for low-pressure mode of excitation (mode 3).*

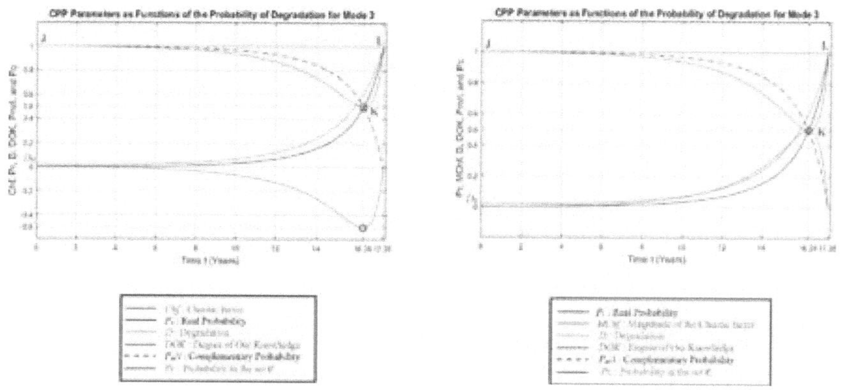


**Figure 23.**
*Degradation and CPP parameters with Chf (a) and with MChf (b) for mode 3.*

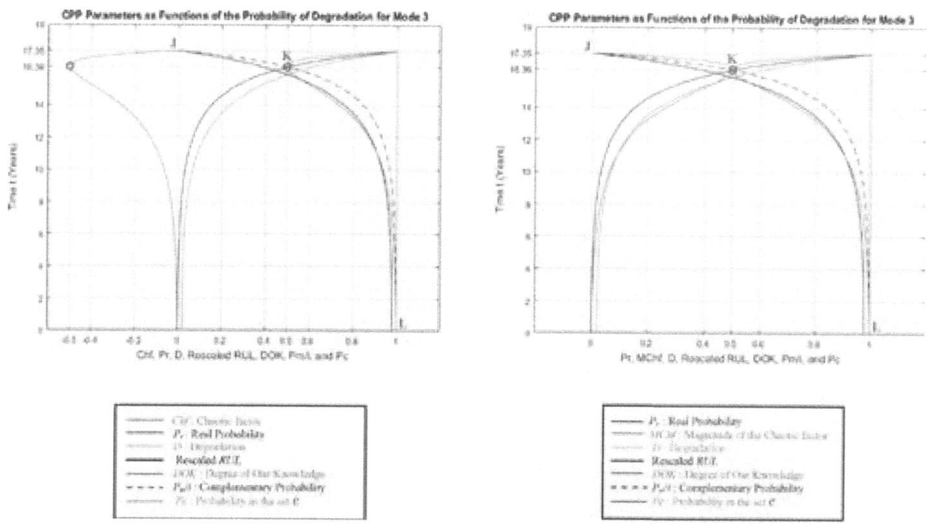


**Figure 24.**
*Degradation, rescaled RUL, and CPP parameters with Chf (a) and with MChf (b) for mode 3.*

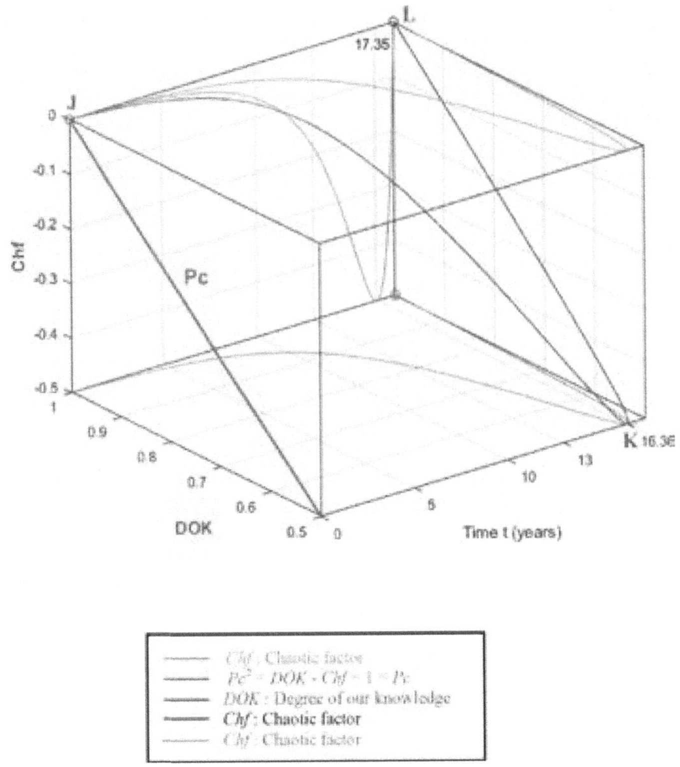


**Figure 25.**
*DOK and Chf in terms of t and of each other for mode 3.*

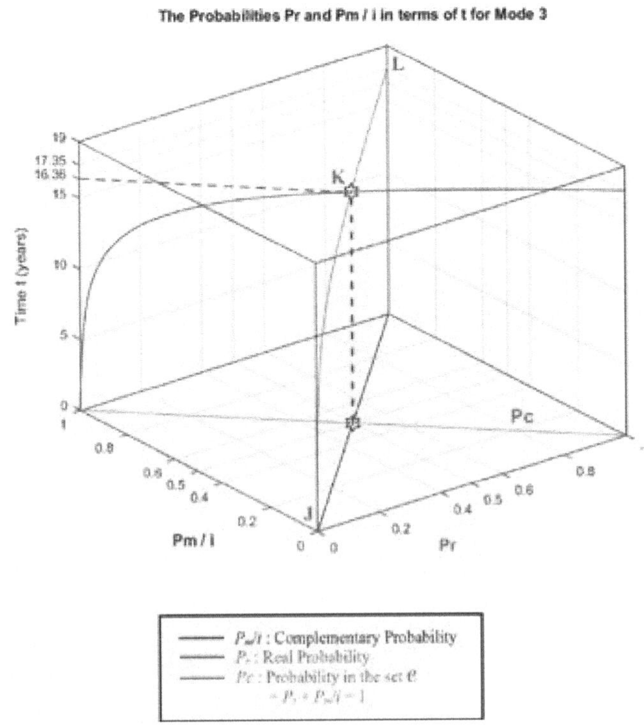


**Figure 26.**
*$P_r$ and $P_m/i$ in terms of t and of each other for mode 3.*

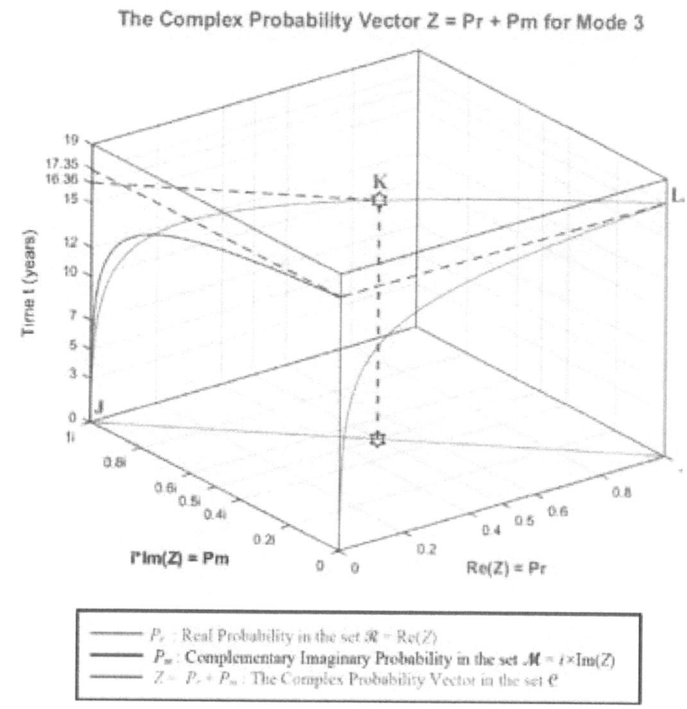


**Figure 27.**
*The complex probability vector Z in terms of t for mode 3.*

## 6. Final analysis: explanation and the general prognostic equations

We will present in this section the original general prognostic equations, we will interpret all the achieved simulations and the obtained data, and we will do a final analysis. Also, we will illustrate the results and a detailed discussion of the all the previous simulations and figures and of the following corresponding tables.

Firstly, we have linked prognostic characterized by the degradation $D(t)$ with probability theory characterized by the *CDF* $F(t)$ by supposing that $D(t) = F(t)$ and the justification for this assumption were given. Consequently, the deterministic $D(t)$ computed from deterministic analytic linear prognostic becomes a nondeterministic cumulative probability distribution function. Therefore, the deterministic and discrete variable of pressure cycles time $t$ becomes a random and discrete variable. Thus, the resultant of all the factors influencing the system which was deterministic becomes a stochastic resultant because $D(t)$ quantifies now the random degradation of the pipeline in terms of the random cycle time $t$. Accordingly, all the parameters' exact values of the $D(t)$ expression (Eq. 6) become now the mean values of the stochastic factors influencing the pipeline and are embodied by *PDFs* as functions of the stochastic variable of pressure cycle time $t$ (refer to Section 3.5). As a matter of fact, this is the real-world case where randomness is omnipresent in one form or another. What we consider and judge as a deterministic phenomenon is nothing in reality but a simplification and an approximation of an actual chaotic and stochastic phenomenon and experiment due to the impact of a huge number of nondeterministic and deterministic forces and factors (a good example is a lottery machine).

Subsequently, we do an updated follow-up of the performance of the random degradation in terms of time or cycle number, which is subject to non-chaotic and chaotic influences, by using the quantity $P_r(t_k)/\psi_j$ due to its definition that evaluates the jumps in the stochastic degradation *CDF* $D(t)$. Hence,

$$P_r(t_k) = \psi_j \times [D(t_k) - D(t_{k-1})], \text{ for any pressure mode } j = 1, 2, 3.$$

Referring to classical probability theory, this makes $P_r(t_k)/\psi_j$ the system probability of failure at $t = t_k$, with $0 \leq P_r(t_k)/\psi_j \leq 1$ and $\sum_{t=t_0}^{t=t_C} P_r(t)/\psi_j$ = [sum of all the jumps in $D$ from $t_0$ to $t_C$] = $D_C$ = 1, just like any probability density function (*PDF*).

In addition, in the simulations, a constant and very small increments in $t$ have been taken which lead to very small increments in $D$ and hence in $P_r(t_k)/\psi_j$. So, we have multiplied those very small jumps in $D$ by a simulation magnifying factor that we called $\psi_j$. Note that $1/\psi_j$ is a normalizing constant that is used to reduce $P_r(t_k)$ function to a probability density function with a total probability equal to one. $1/\psi_j$ is a function of the pressure mode and conditions, and it depends on the parameters in the degradation (Eq. (6)). We have from the simulations $\psi_1 = 5082$ for the high-pressure mode ($j$ = 1, mode 1), $\psi_2 = 6737$ for the middle-pressure mode ($j$ = 2, mode 2), and $\psi_3 = 9151$ for the low-pressure mode ($j$ = 3, mode 3). So we get the following: if $t$ tends to $t_0$ = 0, then $P_r(t_k)$ tends to 0, and if $t$ tends to $t_C$ then $P_r(t_k)$ tends to 1, so $0 \leq P_r(t_k) \leq 1$ and $\sum_{t=t_0}^{t=t_C} P_r(t) = \psi_j \times D_C = \psi_j \times 1 = \psi_j$ as if $P_r(t_k)$ was a *CDF* although mathematically speaking it is not at all. This, since $P_r(t_k)$ is not cumulative, it is just $\psi_j$ times the probability of failure at $t = t_k$. Hence, in the simulations, $P_r(t_k)$ becomes now the probability that the system failure occurs at $t = t_k$ and is used accordingly to compute all the *CPP* parameters.

Therefore, $D(t_k) = F(t_k) = P_{rob}(0 \leq t \leq t_k) = P_{rob}(t = 0 \text{ or } t = 1 \text{ or } t = 2 \text{ or } \ldots \text{ or } t = t_k)$ = sum of all failure probabilities between 0 and $t_k$ = probability that failure will occur somewhere between 0 and $t_k$. So, if $t_k$ = 0 then $P_{rob}(t \leq 0) = D(0) = D_0$

= probability that failure will occur at $t$ = 0 and before. If $t_k = t_C$ then $P_{rob}(0 \leq t \leq t_C) = D(t_C) = 1$ = sum of all failure probabilities between 0 and $t_C$ = probability that failure will occur somewhere between 0 and $t_C$. If $t_k > t_C$ then $P_{rob}(t > t_C) = D(t_C) = 1$ = probability that failure will occur beyond $t_C$. We can see that failure probability increases with the increase of the pressure cycles time $t_k$ until at the end it becomes 1 when $t_k \geq t_C$.

Hence, if $t_0 = 0$ and $D(t_0) = 0$ then

$$D(t_k) = P_{rob}(0 \leq t \leq t_k) = \sum_{t=0}^{t=t_k} P_{rob}(t) = \sum_{t=0}^{t=t_k} P_r(t)/\psi_j$$

This implies that $D(t_C) = P_{rob}(0 \leq t \leq t_C) = \sum_{t=0}^{t=t_C} P_{rob}(t) = \sum_{t=0}^{t=t_C} P_r(t)/\psi_j = 1$ and

$$D(0) = P_{rob}(t \leq 0) = \sum_{t=0}^{t=0} P_{rob}(t) = \sum_{t=0}^{t=0} P_r(t)/\psi_j = P_r(0)/\psi_j = 0.$$

If $t_0 \neq 0$ and $D(t_0) \neq 0$, then the prognostic equation in the new model is

$$D(t_k) = P_{rob}(t_0 \leq t \leq t_k) = \sum_{t=t_0}^{t=t_k} P_{rob}(t) = \sum_{t=t_0}^{t=t_k} P_r(t)/\psi_j \tag{25}$$

for any mode $j$ of pressure profile and with $P_r(t_0)/\psi_j = D_0$.

Moreover, since $P_r(t_k) = \psi_j[D(t_k) - D(t_{k-1})]$, this leads to the following recursive relation:

$$D(t_k) = D(t_{k-1}) + P_r(t_k)/\psi_j; \text{for every } t_k, t_0 \leq t_k \leq t_C. \tag{26}$$

In the case of general prognostic, if we possess the *PDF* of system failure then it can be included in Eqs. (25) and (26) and hence evaluate at any instant $t_k$ the system degradation and vice versa. Consequently, all the other *CPP* model parameters (*DOK*, *Chf*, *MChf*, $P_r$, $P_m$, $P_m/i$, $Z$, $Pc$) will follow. This would be our new prognostic model general equation:

$$D(t_k) = P_{rob}(t_0 \leq t \leq t_k) = \sum_{t=t_0}^{t=t_k} P_{rob}(t) = \sum_{t=t_0}^{t=t_k} PDF_{failure}(t) \tag{27}$$

And the recursive relation

$$D(t_k) = D(t_{k-1}) + PDF_{failure}(t_k) \tag{28}$$

with $PDF_{failure}(t_0) = D_0$.

It is crucial to indicate here that the $PDF_{failure}$ function of the system failure has all the mathematical characteristics and all the possible features of a probability density function whether it is a continuous or a discrete stochastic function and it can follow any imaginable probability distribution in condition only that it characterizes the failure function and the random degradation of the studied system whether it is a petrochemical pipe in the buried, unburied, or offshore case or a vehicle suspension system or any nondeterministic system under the effect of randomness and chaos. In fact, the function $PDF_{failure}$ inherits all the attributes and features of the failure system function and of the nondeterministic degradation.

Furthermore, by applying *CPP* to the pipe prognostic, and in the three simulations of pressure modes, we were successful in the original prognostic model to quantify in $\mathcal{R}$ (our real laboratory) both our chaos embodied by *Chf* and *MChf* and our certain knowledge embodied by *DOK*. These three parameters of *CPP* are evaluated and caused by the resultant of all the nonrandom (deterministic) and random (nondeterministic) aspects influencing the system of pipeline. Knowing that, in the novel paradigm, the factors' resultant effect on *RUL* and *D* is materialized by the jumps in their curves and is accordingly expressed and concretized in $\mathcal{R}$ by $P_r$ and in $\mathcal{M}$ by $P_m$. As it was defined in *CPP*, $\mathcal{M}$ is an imaginary probability extension of the real probability set $\mathcal{R}$, and the complex probability set $\mathcal{C}$ is the sum of both probability sets; thus, $\mathcal{C} = \mathcal{R} + \mathcal{M}$. Because $P_m = i(1 - P_r)$, therefore it is the complementary probability of $P_r$ in $\mathcal{M}$. Hence, if $P_r$ is identified as the failure probability of the system in $\mathcal{R}$ at the pressure cycle time $t = t_k$, then $P_m$ is identified as the corresponding probability in the set $\mathcal{M}$ that the system failure will not occur at the same pressure time $t = t_k$. So, $P_m$ is the associated probability in the set $\mathcal{M}$ of the system survival at $t = t_k$. It follows that $P_m/i = 1 - P_r$ is the associated probability but in the set $\mathcal{R}$ of the system survival at the same pressure cycles time. Accordingly, we know that the sum in $\mathcal{R}$ of both complementary probabilities is surely 1 from classical probability theory. This sum is nothing but $P_C$ which is equal to $P_r + P_m/i = P_r + (1 - P_r) = 1$ always. The sum in $\mathcal{C}$ of both complementary probabilities is the complex random number and vector $Z$ which is equal to $P_r + P_m = P_r + i(1 - P_r)$. And as the complex probability cubes show and illustrate, we realize that $Z$ is the sum in $\mathcal{C}$ of the real probability of failure and of the imaginary probability of survival in the complex probability plane that has the equation

$P_r(t) = iP_m(t) + 1$ for $\forall t : 0 \le t \le t_C$, $\forall P_r : 0 \le P_r \le 1$, and $\forall P_m : 0 \le P_m \le i$. What is interesting is that the square of the norm of $Z$ which is $|Z|^2$ is nothing but *DOK*, as it was proved in *CPP* and in the new model. Moreover, since $MChf = -2iP_rP_m = 2P_rP_m/i$, therefore it is twice the product in $\mathcal{R}$ of both the probability of failure and the probability of survival, and it quantifies the magnitude of chaos since it is always 0 or positive. All the simulations show and prove all these facts.

We can conclude from all the above that since $D(t)$ is a *CDF*, since the factor resultant is random, and since the jumps in $D$ are the simulations failure probabilities $P_r(t_k)$, then we are dealing with a random experiment, thus the natural appearance of *Chf*, *MChf*, *DOK*, *Z*, and hence *Pc*. So, we get in the simulations:

$$Chf(t_k) = -2P_r(t_k)P_m(t_k)/i = -2\Big\{\psi_j[D(t_k) - D(t_{k-1})]\Big\}\Big\{1 - \psi_j[D(t_k) - D(t_{k-1})]\Big\}. \tag{29}$$

$$MChf(t_k) = |Chf(t_k)| = 2\Big\{\psi_j[D(t_k) - D(t_{k-1})]\Big\}\Big\{1 - \psi_j[D(t_k) - D(t_{k-1})]\Big\}. \tag{30}$$

$$\begin{aligned} DOK(t_k) &= 1 - 2P_r(t_k)P_m(t_k)/i \\ &= 1 - 2\Big\{\psi_j[D(t_k) - D(t_{k-1})]\Big\}\Big\{1 - \psi_j[D(t_k) - D(t_{k-1})]\Big\}. \end{aligned} \tag{31}$$

$$Z(t_k) = P_r(t_k) + P_m(t_k) = \psi_j[D(t_k) - D(t_{k-1})] + i\Big\{1 - \psi_j[D(t_k) - D(t_{k-1})]\Big\}. \tag{32}$$

$$Pc^2(t_k) = DOK(t_k) - Chf(t_k) = DOK(t_k) + MChf(t_k) = 1; \text{for every } t_k, 0 \le t_k \le t_C. \tag{33}$$

Furthermore, in the new model, we have

$$RUL(t_k) = t_C - t_k.$$

Note that since $t$ and $D$ are random, then *RUL* is also a random function of $t$. Thus, we have in the set $\mathcal{R}$:

$$\begin{aligned} P_{rob}[RUL(t_k)] &= P_{rob}(\textit{the system will survive for } t_k < t \le t_C) \\ &= 1 - P_{rob}(\textit{the system will fail for } t \le t_k) \\ &= 1 - D(t_k) \\ &= \textit{Rescaled } [RUL(t_k)] \textit{ in all the three pressure modes simulations} \end{aligned} \tag{34}$$

Then, we get always $P_{rob}[RUL(t_k)] + D(t_k) = 1$ everywhere.
This implies that $P_{rob}[RUL(t_k = 0)] = 1 - D(t_k = 0) = 1 - D_0 \approx 1$.
and $P_{rob}[RUL(t_k = t_C)] = 1 - D(t_k = t_C) = 1 - D_C = 1 - 1 = 0$.
Hence, we reach a new and general prognostic equation for *RUL*. If $t_0 \ne 0$ and $D(t_0) \ne 0$ then

$$\begin{aligned} P_{rob}[RUL(t_k)] &= P_{rob}(Survival : t_k < t \le t_C) = 1 - P_{rob}(Failure : t_0 \le t \le t_k) \\ &= 1 - \sum_{t=t_0}^{t=t_k} P_r(t)/\psi_j; \quad \text{with } P_r(t_0)/\psi_j = D_0 \end{aligned} \tag{35}$$

$$\begin{aligned} &= 1 - D(t_k) = \sum_{t=t_{k+1}}^{t=t_C} P_r(t)/\psi_j \\ &= 1 - \sum_{t=t_0}^{t=t_k} PDF_{failure}(t); \quad \text{with } PDF_{failure}(t_0) = D_0 \end{aligned} \tag{36}$$

$$= \sum_{t=t_{k+1}}^{t=t_C} PDF_{failure}(t) \tag{37}$$

for any mode $j$ of pressure profile.

Moreover, from Eqs. (25), (26), (27), and (28) and for any mode $j$ of pressure profile, we have the following recursive relations:

$$P_{rob}[RUL(t_k)] = 1 - D(t_k) = 1 - \left\{D(t_{k-1}) + P_r(t_k)/\psi_j\right\} \tag{38}$$

$$= 1 - \left\{D(t_{k-1}) + PDF_{failure}(t_k)\right\} \tag{39}$$

$$= 1 - \left\{1 - P_{rob}[RUL(t_{k-1})] + P_r(t_k)/\psi_j\right\} \tag{40}$$

$$= P_{rob}[RUL(t_{k-1})] - P_r(t_k)/\psi_j \tag{41}$$

$$= P_{rob}[RUL(t_{k-1})] - PDF_{failure}(t_k) \tag{42}$$

where $P_{rob}[RUL(t_{k-1})] = 1 - D(t_{k-1})$.

In the ideal case, if all the factors are 100% deterministic, then we have in $\mathcal{R}$ the probability of failure for $t_k < t_C$ is 0 and is 1 for $t_k \geq t_C$; accordingly the probability of system survival for $t_k < t_C$ is 1 and is 0 for $t_k \geq t_C$, since certain failure will occur only at $t_k = t_C$. So, degradation is determined surely everywhere in $\mathcal{R}$, and its CDF is replaced by a deterministic function and curve. Therefore, chaos is null, and hence *Chf* = *MChf* = 0, and *DOK* = 1 always for all $0 \leq t_k \leq t_C$. Thus, $P_{rob}[RUL(t_k < t_C)] = 1$ and $P_{rob}[RUL(t_k \geq t_C)] = 0$.

Furthermore, at each instant $t$ in the original prognostic paradigm, the stochastic $RUL(t)$ and $D(t)$ are predicted with certitude in the complex probability set $\mathcal{C}$ with $Pc^2$ = *DOK* – *Chf* = *DOK* + *MChf* maintained as equal to 1 through a continuous compensation between *Chf* and *DOK*. This compensation is from the instant $t$ = 0 where $D(t) = D_0 = 0.020408 \approx 0$ until the instant of failure $t_C$ where $D(t_C) = 1$. Moreover, we can realize that *DOK* does not include any uncertain knowledge (with a probability less than 100%); it is the measure of our certain knowledge (probability = 100%) about the expected event. We can understand that we have eliminated and subtracted in the equation above all the random factors and chaos (*Chf*) from our random experiment when computing $Pc^2$; hence no chaos exists in $\mathcal{C}$, and it only exists (if it does) in $\mathcal{R}$; consequently, this has led to a 100% deterministic outcome and experiment in $\mathcal{C}$ since the probability $Pc$ is constantly equal to 1. This is one of the advantages of extending $\mathcal{R}$ to $\mathcal{M}$ and therefore of working in

| For Any Internal Pressure Mode | $D$ | $P_{rob}[RUL(t)]$ | $DOK$ | $Chf$ | $MChf$ | $P_m/i$ | $Z$ | $Pc$ |
|---|---|---|---|---|---|---|---|---|
| $t=0 \Rightarrow P_r=0$ | $= D_0$ | $= 1-D_0$ | $=1$ | $=0$ | $=0$ | $=1$ | $=i$ | $=1$ |
| $0<P_r<0.5$<br>$t\uparrow \Rightarrow P_r\uparrow$ | ↑ | ↓ | ↓ | ↓ | ↑ | ↓ | Re(Z)↑<br>Im(Z)↓ | $=1$ |
| $t\uparrow$<br>$P_r=0.5$ | ↑ | ↓ | = Min<br>= +0.5 | = Min<br>= -0.5 | = Max<br>= +0.5 | =<br>+0.5 | =<br>0.5+0.5i | $=1$ |
| $0.5<P_r<1$<br>$t\uparrow \Rightarrow P_r\uparrow$ | ↑ | ↓ | ↑ | ↑ | ↓ | ↓ | Re(Z)↑<br>Im(Z)↓ | $=1$ |
| $t=t_C \Rightarrow P_r=1$ | $=1$ | $=0$ | $=1$ | $=0$ | $=0$ | $=0$ | $=1$ | $=1$ |

**Table 2.**
*The new prognostic model parameters for any pipeline internal pressure mode.*

| | $t\uparrow$ and $0 < P_r < 0.5$ | | |
|---|---|---|---|
| Relative Pressure Modes | Mode2 / Mode1 | Mode3 / Mode2 | Mode3 / Mode1 |
| $D$ | ↓ | ↓ | ↓ |
| $P_{rob}[RUL(t)]$ | ↑ | ↑ | ↑ |
| $DOK$ | ↑ | ↑ | ↑ |
| $Chf$ | ↑ | ↑ | ↑ |
| $MChf$ | ↓ | ↓ | ↓ |
| $P_r$ | ↓ | ↓ | ↓ |
| $P_w/i$ | ↑ | ↑ | ↑ |
| $Z$ | Re(Z) ↓<br>Im(Z) ↑ | Re(Z) ↓<br>Im(Z) ↑ | Re(Z) ↓<br>Im(Z) ↑ |
| $Pc$ | ≈1 | ≈1 | ≈1 |

**Table 3.**
*The new prognostic model and the relative pipeline pressure mode comparisons for* $0 < P_r < 0.5$.

| | $t\uparrow$ and $0.5 < P_r < 1$ | | |
|---|---|---|---|
| Relative Pressure Modes | Mode2 / Mode1 | Mode3 / Mode2 | Mode3 / Mode1 |
| $D$ | ↓ | ↓ | ↓ |
| $P_{rob}[RUL(t)]$ | ↑ | ↑ | ↑ |
| $DOK$ | ↓ | ↓ | ↓ |
| $Chf$ | ↓ | ↓ | ↓ |
| $MChf$ | ↑ | ↑ | ↑ |
| $P_r$ | ↓ | ↓ | ↓ |
| $P_w/i$ | ↑ | ↑ | ↑ |
| $Z$ | Re(Z) ↓<br>Im(Z) ↑ | Re(Z) ↓<br>Im(Z) ↑ | Re(Z) ↓<br>Im(Z) ↑ |
| $Pc$ | ≈1 | ≈1 | ≈1 |

**Table 4.**
*The new prognostic model and the relative pipeline pressure mode comparisons for* $0.5 < P_r < 1$.

$\mathcal{C} = \mathcal{R} + \mathcal{M}$. Thus, in the original prognostic paradigm, our knowledge of all the indicators and parameters ($RUL$, $P_{rob}$, $D$, etc.) is totally predictable, always perfect, and constantly complete because $Pc$ = 1 permanently, independently of any random factors or any pressure profile (**Table 2**).

Finally, we say that we have applied for pressure modes 2 and 3 the same analysis, logic, and methodology that we have used for pressure mode 1 regarding the remaining useful lifetime, the degradation, as well as all the *CPP* parameters (**Tables 3** and **4**). Therefore, we can accordingly infer that whatever the pressure conditions and environment are, then the results and conclusions are analogous. This demonstrates the strength and soundness of the novel axioms adopted and of the new prognostic paradigm developed.

## 7. Conclusion and perspectives

The high availability of technological systems, like defense, aerospace, automobile industries, and petrochemistry, is a central major objective of previous and latest developments in the technology of system design where it is very well-known that expensive failure may in general happen unexpectedly. A novel model of analytic prognostic was established in my earlier work and publications as a counterpart of existent classical strategies of maintenance in order to take into account the evolving environment and product state and in order to make them more efficient. We have applied this model to systems of petrochemical pipes that are exposed to fatigue failure under cyclic repetitive triangular pressure. It is known that the effects of fatigue will initiate micro-cracks that can spread rapidly and hence will lead to failure. This model is founded on existing laws of damage in fracture mechanics which are the law of Palmgren-Miner of linear damage accumulation and the law of Paris-Erdogan of crack propagation. This prognostic model estimates the system *RUL* from a predefined threshold of degradation $D_C$. The model of degradation established in this earlier work is founded on the damage measurement $D$ accumulation after each cycle time of pressure. The system is judged to be in wear-out state when this measured and predefined threshold $D_C$ is reached. Moreover, to make the model more realistic and accurate, we have taken into consideration the stochastic influences afterward as well here. We have applied this model to the industry of pipelines; therefore, a prognostic study of the pipeline system enables us to enhance its strategies of maintenance.

In the present research work, the novel extended Kolmogorov paradigm of eight axioms (*EKA*) was applied and bonded to the analytic and linear prognostic of buried petrochemical pipeline systems subject to fatigue. Hence, a tight link between the remaining useful lifetime or degradation and the original paradigm was made. Therefore, the model of "complex probability" was more elaborated beyond the scope of my previous 12 research works on this subject.

Although the analytic linear laws of prognostic are very well-known and deterministic in [14, 16], there are general influences and aspects that can be chaotic and stochastic (like humidity, temperature, material nature, geometry dimensions, applied load location, water action, corrosion, soil pressure and friction, atmospheric pressure, etc.). Moreover, various variables in the expressions (5) and (6) of degradation which are considered as deterministic can also have a random aspect, such as the magnitude of applied pressure (due to the different conditions of pressure profile) and the length of the initial crack (potentially existing from the process of manufacturing). All those stochastic factors, embodied in the model by their mean values, influence the buried pipeline system and make its function of degradation diverge from its computed trajectory modeled by these deterministic laws. An updated follow-up of the degradation performance and behavior with cycle number or time, which is subject to non-chaotic and chaotic influences, is made possible by $P_r(t_k)/\psi_j$ due to its definition that evaluates the jumps in $D$. In fact, chaos modifies and affects all the environment and system parameters included in the degradation equations (Eqs. (5) and (6)). Consequently, chaos total effect on the pipelines contributes to shape the degradation curve $D$ and is materialized by and counted in the pipeline system failure probability $P_r(t_k)/\psi_j$. Actually, $P_r(t_k)/\psi_j$ quantifies the resultant of all the nonrandom (deterministic) and random (nondeterministic) parameters and aspects which are contained in the equation of $D$, which affect the system and which lead to the consequent final curve of degradation. Consequently, an accentuated influence of chaos on the pipeline can lead to a smaller (or bigger) jump in the trajectory of degradation and therefore to a smaller (or bigger) failure probability $P_r(t_k)/\psi_j$.

Additionally, as it was verified and shown in the novel model, when the degradation index is 0 or 1 and correspondingly the *RUL* is $t_C$ or 0, then the chaotic factor (*Chf* and *MChf*) is zero, and the degree of our knowledge (*DOK*) is 1 since the system state is totally known. During the process of degradation ($0 < D < 1$), we have $-0.5 \leq Chf < 0$, $0 < MChf \leq 0.5$, and $0.5 \leq DOK < 1$. Notice that during this whole process, we have always $Pc^2 = DOK - Chf = DOK + MChf = 1 = Pc$, which means that the phenomenon which looked to be stochastic and random in the set $\mathcal{R}$ is now certain and deterministic in the set $\mathcal{C} = \mathcal{R} + \mathcal{M}$, and this after the addition of the contributions of $\mathcal{M}$ to the phenomenon occurring in $\mathcal{R}$ and thus after subtracting and eliminating the chaotic factor from the degree of our knowledge. Moreover, the probabilities of the system survival and of failure corresponding to each instant $t$ have been evaluated, in addition to the probability of *RUL* after a pressure cycles time $t$, which are all functions of the stochastic degradation jump. Consequently, at each instance of $t$, all the novel *CPP* parameters $D$, $RUL$, $P_r$, $P_m$, $P_m/i$, $DOK$, $Chf$, $MChf$, $Pc$, and $Z$ are certainly and perfectly predicted in the complex probability set $\mathcal{C}$ with $Pc$ maintained as equal to 1 constantly and permanently. Furthermore, using all these illustrated simulations and drawn graphs all over the whole research work, we can quantify and visualize both the certain knowledge (expressed by *DOK* and *Pc*) and the system chaos and random effects (expressed by *Chf* and *MChf*) of the pipeline system. This is definitely very fascinating, fruitful, and wonderful and proves once again the advantages of extending the five probability axioms of Kolmogorov and thus the novelty and benefits of this original field in prognostic and applied mathematics that can be called verily "The Complex Probability Paradigm."

As a prospective and future work and challenges, and concerning some applications to practical engineering, it is planned to more elaborate the original created prognostic paradigm and to implement it to a varied set of nondeterministic and dynamic systems like vehicle suspension systems and offshore and buried petrochemical pipes which are under the influence of fatigue and in the cases of nonlinear and linear damage accumulation. Furthermore, we will apply also *CPP* to other random experiments in classical probability theory and in stochastic processes and to the field of prognostic in engineering using the first order reliability method (FORM) as well as to the random walk problems which have enormous applications in physics, in economics, in chemistry, in applied and pure mathematics.

## Nomenclature

| | |
|---|---|
| $\mathcal{R}$ | The set of real probabilities of events |
| $\mathcal{M}$ | The set of imaginary probabilities of events |
| $\mathcal{C}$ | The set of complex probabilities of events |
| $i$ | The imaginary number where $i^2 = -1$ and $i = \sqrt{-1}$ |
| *EKA* | extended Kolmogorov axioms |
| *CPP* | Complex probability paradigm |
| $P_{rob}$ | any event probability |
| $P_r$ | system failure probability, probability in the real set $\mathcal{R}$ |
| $P_m$ | system survival probability in $\mathcal{M}$, probability in the imaginary set $\mathcal{M}$ corresponding to the real probability in $\mathcal{R}$ |
| $P_m/i$ | system survival probability in $\mathcal{R}$ |
| $Pc$ | probability in the complex set $\mathcal{C}$, probability of an event in $\mathcal{R}$ with its associated event in $\mathcal{M}$ |

| | |
|---|---|
| $Z$ | the sum of $P_r$ and $P_m$, complex probability number and vector |
| $DOK$ | $= \|Z\|^2$, the square of the norm of $Z$, degree of our knowledge of the random event and experiment |
| $Chf$ | chaotic factor |
| $MChf$ | magnitude of the chaotic factor |
| $t$ | pressure cycle time |
| $t_C$ | pressure cycle time till system failure |
| $P_j$ | pipelines internal triangular pressure |
| $f_j(t)$ | probability density function for each pressure mode $j$ |
| $F(t)$ | cumulative probability distribution function |
| $\psi_j$ | simulation magnifying factors for each pressure mode $j$ |
| $1/\psi_j$ | the normalizing constant of $P_r(t)$ for each pressure mode $j$ |
| $D$ | degradation indicator of a system |
| $RUL$ | remaining useful lifetime of a system |
| $P_{rob}[RUL(t)]$ | probability of $RUL$ after a pressure cycle time $t$. |

## References


[1] El-Tawil K, Kadry S, Abou Jaoude A. Life time estimation under probabilistic fatigue of cracked plates for multiple limits states. In: International Conference (ICNAAM 2009) on Numerical Analysis and Applied Mathematics, Rethymno, Crete, Greece; September 18–22. 2009

[2] Abou Jaoude A, El-Tawil K, Kadry S, Noura H, Ouladsine M. Analytic prognostic model for a dynamic system. International Review of Automatic Control (IREACO). 2010; **3**(6):568-577

[3] Abou Jaoude A. Analytic and linear prognostic model for a vehicle suspension system subject to fatigue. Systems Science & Control Engineering (SSCE). 2015;**3**(1):81-98

[4] Abou Jaoude A, El-Tawil K. Analytic and nonlinear prognostic for vehicle suspension systems. American Journal of Engineering and Applied Sciences (AJEAS). 2013;**6**(1):42-56

[5] El-Tawil K, Abou Jaoude A, Kadry S, Noura H, Ouladsine M. Prognostic based on analytic laws applied to petrochemical pipelines. In: International Conference on Computer-Aided Manufacturing and Design (CMD 2010); China. November 2010

[6] Abou Jaoude A, Kadry S, El-Tawil K, Noura H, Ouladsine M. Analytic prognostic for petrochemical pipelines. Journal of Mechanical Engineering Research (JMER). April 2011;**3**(3):64-74

[7] Vachtsevanos G, Lewis F, Roemer M, Hess A, Wu B. Intelligent Fault Diagnosis and Prognosis for Engineering Systems. John Wiley & Sons, Inc.; 2006

[8] Lemaitre J, Chaboche J. Mechanics of Solid Materials. New York: Cambridge University Press; 1990

[9] Vasile OE. Contribution Au Pronostic De Défaillances Par Réseau Neuro-Flou : Maitrise De L'erreur De Prédiction [Thèse de doctorat]. 2008

[10] Peysson F, Ouladsine M, Outbib R, Leger J-B, Myx O, Allemand C. A generic prognostic methodology using damage trajectory models. IEEE Transactions on Reliability. 2009;**58**:277-285

[11] PROTEUS WP2 Team. On the Use of Artificial Intelligence for Prognosis and Diagnosis in the PROTEUS E-Maintenance Platform. ITEA European Project; 2005

[12] Concha ME. Fault diagnostic and failure prognostic in nonlinear dynamic system. IEEE Transactions on Control System Technology. 2007

[13] Abou Jaoude A, El-Tawil K, Kadry S, Noura H, Ouladsine M. Prognostic model for buried tubes. International Conference on Advanced Research and Applications in Mechanical Engineering (ICARAME'11), Notre Dame University-Louaizé, Lebanon, June 13–15, 2011

[14] Abou Jaoude A, Noura H, El-Tawil K, Kadry S, Ouladsine M. Lifetime analytic prognostic for petrochemical pipes subject to fatigue, SAFEPROCESS. In: 8th IFAC Symposium on Fault Detection, Supervision and Safety of Technical Processes, Mexico City, Mexico, August 29–31. 2012

[15] Abou Jaoude A, Noura H, El-Tawil K, Kadry S, Ouladsine M. Analytic prognostic model for stochastic fatigue of petrochemical pipelines. In: Australian Control Conference (AUCC 2012), Sydney, Australia, November. 2012. pp. 15-16

[16] Abou Jaoude A. Advanced analytical model for the prognostic of industrial systems subject to fatigue [PhD thesis].

Aix-Marseille Université and the Lebanese University, defended on December 7; 2012

[17] El-Tawil K, Abou Jaoude A. Stochastic and nonlinear based prognostic model. Systems Science & Control Engineering (SSCE). 2013;**1**(1):66-81

[18] Abou Jaoude A, El-Tawil K. Stochastic prognostic paradigm for petrochemical pipelines subject to fatigue. American Journal of Engineering and Applied Sciences (AJEAS). 2013;**6**(2):145-160

[19] Abou Jaoude A. Automatic Control and Prognostic. Saarbrucken, Germany: Scholars' Press; 2013

[20] Abou Jaoude A, El-Tawil K, Kadry S. Prediction in complex dimension using Kolmogorov's set of axioms. Journal of Mathematics and Statistics (JMSS). 2010;**6**(2):116-124

[21] Abou Jaoude A. The complex statistics paradigm and the law of large numbers. Journal of Mathematics and Statistics (JMSS). 2013;**9**(4):289-304

[22] Abou Jaoude A. The theory of complex probability and the first order reliability method. Journal of Mathematics and Statistics (JMSS). 2013;**9**(4):310-324

[23] Abou Jaoude A. Complex probability theory and prognostic. Journal of Mathematics and Statistics (JMSS). 2014;**10**(1):1-24

[24] Abou Jaoude A. The complex probability paradigm and analytic linear prognostic for vehicle suspension systems. American Journal of Engineering and Applied Sciences (AJEAS). April 2015;**8**(1):147-175

[25] Abou Jaoude A. The paradigm of complex probability and the Brownian motion. Systems Science and Control Engineering (SSCE). 2015;**3**(1):478-503

[26] Abou Jaoude A. The paradigm of complex probability and Chebyshev's inequality. Systems Science and Control Engineering (SSCE). 2016;**4**(1):99-137

[27] Abou Jaoude A. The paradigm of complex probability and analytic nonlinear prognostic for vehicle suspension systems. Systems Science and Control Engineering (SSCE). 2016; **4**(1):99-137

[28] Abou Jaoude A. The paradigm of complex probability and analytic linear prognostic for unburied petrochemical pipelines. Systems Science and Control Engineering (SSCE). 2017;**5**(1):178-214

[29] Abou Jaoude A. The paradigm of complex probability and Claude Shannon's information theory. Systems Science and Control Engineering (SSCE). 2017;**5**(1):380-425

[30] Abou Jaoude A. The paradigm of complex probability and analytic nonlinear prognostic for unburied petrochemical pipelines. Systems Science and Control Engineering (SSCE). 2017;**5**(1):495-534

[31] Abou Jaoude A. The paradigm of complex probability and Ludwig Boltzmann's entropy. Systems Science and Control Engineering (SSCE). 2018; **6**(1):108-149

[32] Chan Man Fong CF, De Kee D, Kaloni PN. Advanced Mathematics for Applied and Pure Sciences. Amsterdam, The Netherlands: Gordon and Breach Science Publishers; 1997

[33] Abou Jaoude A. Applied mathematics: Numerical methods and algorithms for applied mathematicians (PhD thesis). Bircham International University; 2004. Available from: http://www.bircham.edu

[34] Abou Jaoude A. Computer science: Computer Simulation of Monté Carlo methods and random phenomena (PhD thesis). Bircham International University; 2005. Available from: http://www.bircham.edu

[35] Abou Jaoude A. Applied statistics and probability: Analysis and algorithms for the statistical and stochastic paradigm [PhD thesis]. Bircham International University; April 2007. Available from: http://www.bircham.edu

[36] Stepić AI, Ognjanović Z. Complex valued probability logics. Nouvelle Série. 2014;**95**(109):73-86. DOI: 10.2298/PIM1409073I

[37] Cox DR. A use of complex probabilities in the theory of stochastic processes. Mathematical Proceedings of the Cambridge Philosophical Society. 1955;**51**:313-319

[38] Weingarten D. Complex probabilities on $R^N$ as real probabilities on $C^N$ and an application to path integrals. Physical Review Letters. 2002;**89**. DOI: 10.1103/PhysRevLett.89.240201

[39] Youssef S. Quantum mechanics as complex probability theory. Modern Physics Letters A. 1994;**9**:2571-2586

[40] Fagin R, Halpern J, Megiddo N. A logic for reasoning about probabilities. Information and Computation. 1990;**87**:78-128

[41] Bidabad B. Complex probability and Markov stochastic processes. In: Proceedings of the First Iranian Statistics Conference. Tehran: Isfahan University of Technology; 1992

[42] Ognjanović Z, Marković Z, Rašković M, Doder D, Perović A. A probabilistic temporal logic that can model reasoning about evidence. Annals of Mathematics and Artificial Intelligence. 2012;**65**:1-24

[43] Christensen RM. A Physically Based Cumulative Damage Formalism. Stanford, CA: Lawrence Livermore National Laboratory and Stanford University; 2007

[44] Sankavaram C, Kodali A, Ayala DFM, Pattipati K, Singh S, Bandyopadhyay P. Model-based and data-driven prognosis of automotive and electronic systems. In: 5th Annual IEEE Conference on Automation Science and Engineering, Bangalore, India, August 22–25. 2009

[45] Husin Z, Rahman MM, Kadirgama K, Noor MM, Bakar RA. Prediction of fatigue life on lower suspension arm subjected to variable amplitude loading. In: National Conference in Mechanical Engineering Research and Postgraduate Studies, 2nd NCMER 2010, Malaysia, December. 2010. pp. 100-116

[46] Beden SM, Abdullah S, Ariffin AK. Review of fatigue crack propagation models for metallic components. European Journal of Scientific Research. 2009;**28**(3):364-397

[47] Huang C. Structural Health Monitoring System for Deepwater Risers with Vortex-Induced Vibration: Nonlinear Modeling, Blind Identification Fatigue/Damage Estimation and Local Monitoring Using Magnetic Flux Leakage [a thesis submitted in partial fulfillment of the requirements for the Degree Doctor of Philosophy]. Houston, Texas: Mechanical Engineering and Material Science; June 2012

[48] Guan X, Jha R, Liu Y. Trans-dimensional MCMC for fatigue prognosis model determination, updating, and averaging. In: Annual Conference of the Prognostics and Health Management Society (PHM), Portland, USA, October 13–16. 2010

[49] Xiang Y, Liu Y. Efficient probabilistic methods for real-time

fatigue damage prognosis. In: Annual Conference of the Prognostics and Health Management Society (PHM), Portland, USA, October 13–16. 2010

[50] Wei Y, Qiu J, Karimi HR, Wang M. New results on H∞ dynamic output feedback control for Markovian jump systems with time-varying delay and defective mode information. Optimal Control, Applications and Methods. November 2014;**35**(6):656-675

[51] Wei Y, Qiu J, Karimi HR, Wang M. H∞ model reduction for continuous-time Markovian jump systems with incomplete statistics of mode information. International Journal of Systems Science. September 2014;**45**(7): 1496-1507

[52] Wei Y, Qiu J, Karimi HR. Quantized H∞ filtering for continuous-time Markovian jump systems with deficient mode information. Asian Journal of Control. September 2015;**17**(5): 1914-1923

[53] Wei Y, Peng X, Qiu J. Robust and non-fragile static output feedback control for continuous-time semi-Markovian jump systems. Transactions of the Institute of Measurement and Control. September 2016;**38**(9): 1136-1150

Chapter 6

# Fault Detection of Single and Interval Valued Data Using Statistical Process Monitoring Techniques


## Abstract

Principal component analysis (PCA) is a linear data analysis technique widely used for fault detection and isolation, data modeling, and noise filtration. PCA may be combined with statistical hypothesis testing methods, such as the generalized likelihood ratio (GLR) technique in order to detect faults. GLR functions by using the concept of maximum likelihood estimation (MLE) in order to maximize the detection rate for a fixed false alarm rate. The benchmark Tennessee Eastman Process (TEP) is used to examine the performance of the different techniques, and the results show that for processes that experience both shifts in the mean and/or variance, the best performance is achieved by independently monitoring the mean and variance using two separate GLR charts, rather than simultaneously monitoring them using a single chart. Moreover, single-valued data can be aggregated into interval form in order to provide a more robust model with improved fault detection performance using PCA and GLR. The TEP example is used once more in order to demonstrate the effectiveness of using of interval-valued data over single-valued data.



**Keywords:** principal component analysis, generalized likelihood ratio, hypothesis testing, fault detection, Tennessee Eastman Process, interval data


## 1. Introduction

Current technological advancements allow data to be collected from a number of different sources. The availability of abundant data collected from different sensors is beneficial, as they can be utilized in order to observe trends between and within different measured process variables. This allows process models to be developed in order to help identify if different processes or applications are behaving as expected [1]. Additionally, with industrial growth present in many developing countries, efficient process monitoring is essential for newer and more complex processes. Monitoring of these processes is required in order to ensure process safety, maintain product quality, increase economic benefits, and also to ensure that the process adheres to strict environmental regulation standards [2].

Statistical process monitoring methods can be classified into three broad categories: quantitative model based methods, qualitative model based methods, and process history based methods [3–5]. Quantitative model based methods require detailed knowledge of a process in order to construct a model that can be used for monitoring, for example, Kalman filters [3], while qualitative model based methods require the presence of process engineering experts in order to develop monitoring procedures or tasks, for example, fault trees [4]. In the absence of these two requirements, and due to the complexity of many processes that require monitoring, data-based techniques are often commonly used by the industry for various applications from drug design, to drinking water treatment [5–7].

Principal component analysis (PCA) is a powerful, linear data analysis technique widely used in research and industrial applications [8], for fault detection and isolation, data modeling and reconstruction, feature extraction, and noise filtration. PCA is useful for the extraction of dominant underlying information from a dataset, without any previous knowledge of the model. An example of the practical application of PCA has been discussed in [8], where data gathered from parallel sensors are used to quantify the quality of a given food sample. PCA is used to reduce the dimensionality of a dataset, whilst filtering out variability caused by noise [9]. The PCA model has been utilized in order to monitor a wide variety of processes, and has seen many extensions [10–13]. Two main fault detection statistics are typically utilized with a PCA model: Hotelling's $T^2$ statistic, and the Q statistic [10]. Variations captured by the principal component space are monitored using the $T^2$ statistic, while variations in the residual space are monitored using the Q statistic [14].

On the other hand, statistical hypothesis testing methods function by using statistical techniques in order to determine if observations collected from a given process follow the null hypothesis, that is, operating under normal operating conditions, or alternate hypothesis, that is, operating under abhorrent or faulty operating conditions [15]. These faults can be of different types, such as shifts in the mean, variance, or both. The generalized likelihood ratio (GLR) technique has received a lot of attention in process monitoring literature [10, 11, 13, 16]. The GLR method aims to maximize the detection rate for a fixed false alarm rate [15]. Therefore, an objective of this work is to provide a comparative review of the different GLR charts by utilizing examples such as the benchmark Tennessee Eastman Process (TEP) [17].

Data utilized in the construction of a PCA model may be of two types depending on the application being monitored: single-valued, and interval-valued. Single-valued data can be directly obtained from sensors measuring particular variables in a process, while interval-valued data is aggregated or artificially generated from batch single-valued measurements, thereby resulting in a range of possible measurement values for a given process variable at one time instant. The use of interval data in fault detection was originally introduced in order to reduce large datasets to a more manageable size [18], without compromising the integrity of the dataset. In addition, the use of interval data is beneficial because of its inherent ability to deal with missing values in samples, which may happen due to malfunctioning sensors or varying sampling frequencies between variables [19].

However, in cases where reducing the dataset may not be a viable option, due to a relatively limited sample size or sampling frequency, the use of interval data can be applied using a moving window aggregation method. This is also true of applications where batch process monitoring is not a viable option, thereby necessitating the need for real-time online monitoring of samples. The benchmark TEP example will be used once more in order to analyze the benefit of using moving window interval aggregation on the fault detection performance of PCA and GLR.

The rest of this chapter will be organized as follows. In Section 2, a more detailed introduction to PCA is provided along with a quick overview of the fault detection statistics used to examine the fault detection performance of the methods discussed in this paper. Section 3 will introduce hypothesis testing methods and the different GLR charts. In Section 4, the moving window interval aggregation method is explained, as well as its integration with PCA and GLR for the purposes of fault detection. Section 5 then presents illustrative examples using simulated synthetic data and TEP using a PCA-based GLR technique, used to demonstrate the effect that using GLR and interval data has on the fault detection performance. Conclusions are then presented in Section 6.

## 2. Principal component analysis (PCA)

Principal component analysis (PCA) is a linear dimensionality reduction tool used to reduce the number of variables in a dataset, whilst retaining most of the data's variability. PCA finds a new set of variables, called principal components, using a linear combination of the dataset's original cross-correlated variables [9]. The algorithm for PCA is summarized below.

### 2.1 PCA algorithm

Given a $n \times p$ classical training dataset $X$, where $n$ is the number of sample rows and $p$ is the number of variable columns, the PCA model is found as follows:

1. Find the correlation matrix $R$ of $X$.

2. Find the column eigenvectors matrix $P$ and the diagonal eigenvalues matrix $\Lambda$ of $R$. Each eigenvector defines the linear combination coefficients used to find the principal components from the original variables, and each eigenvalue represents the amount of variance that its respective principal component covers in the dataset.

3. Retain $l$ principal components that cover the minimum desired variability in the dataset, denoted as $\hat{P}$.

4. Find the predictive transformation matrix, $\hat{C} = \hat{P}\hat{P}^T$.

5. Find the residual transformation matrix, $\tilde{C} = 1 - \hat{C}$.

$\hat{C}$ is used to find the projection of the dataset onto the PCA model, and $\tilde{C}$ is used to find the amount of deviation of the dataset from its projection onto the PCA model, also known as the matrix of residuals. For more comprehensive details, please refer to [9, 19, 20].

The training dataset $X$ defines the system under normal or optimal operating conditions, where there are no faults and the noise is minimal. Consequently, $X$ is used to find the PCA model, defined using $\hat{C}$ and $\tilde{C}$ transformation matrices. The testing dataset $S$ defines the system under unknown operating conditions, and it is monitored for faults using its respective residuals $\tilde{S} = S \cdot \tilde{C}$, as will be discussed later.

### 2.2 Fault detection statistics

Knowing the optimal number of eigenvectors or principal components to retain, fault detection is then carried out by evaluating the PCA model's residuals using any detection statistic. This section will focus on briefly introducing the two most well-known statistics in literature: The Q and $T^2$ statistics.

The Q-statistics of a $n \times p$ classical residual matrix $\tilde{X}$ is defined as [11]:

$$Q_x[i] = \sum_{j=1}^{p} \left(\tilde{X}_j[i]\right)^2 \tag{1}$$

$Q_x$ is used to find the Q-threshold value $\gamma$, which defines the maximum possible value for a testing data's Q-statistic, denoted as $Q_s$, beyond which the sample will be declared as a fault [14, 19, 21]. The threshold is calculated using the empirical cumulative distribution function (CDF) of $Q_x$, which is an estimate of the true CDF of its discrete values.

The fault detection performance is tabulated by comparing $Q_s$ with $\gamma$. If $Q_s[i]>\gamma$, then the $i$th sample is declared as faulty, otherwise it is normal. There are two metrics used for benchmarking each method: false alarm rate (FAR) and detection rate (DR).

FAR is the average percentage of samples that were wrongfully declared as faults. The detection rate is the average percentage of samples that were rightfully declared as faults. It is desirable to maximize DR, for a fixed FAR, in order to have a better fault detector.

Alternatively, the Hotelling $T^2$ statistic, which measures variations in the principal component space can be used, is computed as follows [22]:

$$T^2 = x^T \hat{P} \hat{\Lambda}^{-1} \hat{P}^T x, \tag{2}$$

where, $\hat{\Lambda} = \text{diag}(\lambda_1, \lambda_2, \ldots, \lambda_l)$, is a diagonal matrix that contains the eigenvalues that are associated with the $l$ retained principal components The threshold for the $T^2$ statistic can be computed either computational or empirically [22]. The Q statistic is often utilized by authors instead of the $T^2$ statistic as it better able to detect smaller faults [10, 11].

## 3. Hypothesis testing methods

Hypothesis testing methods such as the generalized likelihood ratio (GLR), have received a lot of attention in recent literature [10, 13, 23]. Hypothesis testing methods utilize fundamental statistical theory in order to determine if given data conforms to a targeted distribution, that is, a null hypothesis, or deviates from this distribution, and follows an alternative distribution, that is, an alternate hypothesis [15]. In process monitoring terms, the parameters of the null and alternate hypotheses are defined using data from normal and abhorrent operating conditions, respectively [1].

### 3.1 Generalized likelihood ratio

The generalized likelihood ratio (GLR) technique defines the alternate hypotheses by parameters that can assume an infinite number of values, and is therefore called a composite hypothesis. An efficient point estimation method that utilizes the concept of maximum likelihood estimates (MLEs) is employed in order to estimate the required parameters.

The univariate GLR chart uses the concept of maximum likelihood estimates in order to maximize the detection rate for a fixed false alarm rate. The GLR process is accomplished through the following steps [15]:

1. The null and alternate hypotheses are defined, and their respective likelihood functions are derived.

2. Any unknown parameters in the alternate hypothesis are computed from the testing data using their MLEs, for example, the mean and/or variance.

3. The log likelihood ratio of the alternate to null hypotheses is then computed, and its maximum value is calculated, which maximizes the detection rate.

Univariate GLR charts can be designed based on the type of the fault that needs to be detected. Most processes experience shifts in the mean, and/or shifts in the variance, and three of these GLR charts will be explained next.

For the case when residuals are collected from processes under normal operating conditions, the likelihood function derived from a random normal distribution can be defined as follows [24]:

$$L\left(\infty, \mu_0, \sigma_0^2 | x_1, x_2, \ldots, x_k\right) = (2\pi)^{-k/2}\left(\sigma_0^2\right)^{-k/2} \exp\left(-\frac{1}{2\sigma_0^2}\sum_{i=1}^{k}(x_i - \mu_0)^2\right) \quad (3)$$

where $\mu_0$ and $\sigma_0^2$ mean and variance of the process variable measured under normal operating conditions respectively.

### *3.1.1 Univariate GLR chart for a shift in the mean*

If a shift in the mean has occurred at time $\tau$, from $\mu_0$ to $\mu_1$, the likelihood function of the alternate hypothesis is defined as follows [24]:

$$\begin{aligned} &L\left(\tau, \mu_1, \sigma_0^2 | x_1, x_2, \ldots, x_k\right) \\ &= (2\pi)^{-k/2}\left(\sigma_0^2\right)^{-k/2} \exp\left(-\frac{1}{2\sigma_0^2}\left(\sum_{i=1}^{\tau}(x_i - \mu_0)^2 + \sum_{i=\tau+1}^{k}(x_i - \mu_1)^2\right)\right) \end{aligned} \quad (4)$$

Since the magnitude of the new mean is unknown, its MLE can be computed using testing data as follow [24]:

$$\hat{\mu}_{1,\tau,k} = \frac{1}{(k-\tau)}\sum_{i=\tau+1}^{k} x_i. \quad (5)$$

The GLR statistic designed to specifically monitor a shift in the mean can now be computed by taking the log-likelihood ratio of (Eqs. (3) and (4)) [24]:

$$R_k = \max_{0 \le \tau < k} \frac{(k-\tau)}{2\sigma_0^2}\left(\hat{\mu}_{1,\tau,k} - \mu_0\right)^2. \quad (6)$$

The authors in [24] state that it is not necessary to store the entire length of previous historical data in order to compute the MLEs, but a window length of about 400 is sufficient to provide reliable results. Therefore, a window length of 400 was utilized throughout this work for all GLR charts.

### 3.1.2 Univariate GLR chart for a shift in the variance

If only a shift in the variance has occurred from at time $\tau$, from $\sigma_0^2$ to $\sigma_1^2$, the alternate hypothesis for this case is defined as follows [25]:

$$L\left(\tau, \mu_0, \sigma_1^2 | x_{\tau+1}, x_2, \ldots, x_k\right) = (2\pi)^{-k/2}\left(\sigma_1^2\right)^{-k/2} \exp\left(-\frac{1}{2\sigma_1^2}\left(\sum_{i=\tau+1}^{k}(x_i - \mu_0)^2\right)\right). \tag{7}$$

From a quality control standpoint we are only concerned with increases in variance, as larger variations imply that product is being manufactured with quality further away from the targeted amount, and since the magnitude of the new variance is unknown, its MLE can be computed using testing data as follows [25]:

$$\hat{\sigma}_{1,\tau,k}^2 = \max\left\{\sigma_0^2, \frac{1}{k-\tau}\sum_{i=\tau+1}^{k}(x_i - \mu_0)^2\right\}. \tag{8}$$

The GLR statistic designed to specifically monitor a shift in the variance can now be computed by taking the log-likelihood ratio of (Eqs. (3) and (7)) [25]:

$$R_k = \max_{0 \le \tau < k} \frac{k-\tau}{2}\left[\frac{\hat{\sigma}_{1,\tau,k}^2}{\sigma_0^2} - 1 - \ln\left(\frac{\hat{\sigma}_{1,\tau,k}^2}{\sigma_0^2}\right)\right] \tag{9}$$

### 3.1.3 Univariate GLR chart for a shift in the mean and/or variance

Since it is possible for most processes to experience both shifts in the mean and variance, a GLR statistic that is capable of detecting either type of shift can be designed. The likelihood function of the alternate hypothesis for this case is defined as follows [26]:

$$L\left(\tau, \mu_1, \sigma_1^2 | x_1, x_2, \ldots, x_k\right) = (2\pi)^{-k/2}\left(\sigma_0^2\right)^{-\tau/2}\left(\sigma_1^2\right)^{-(k-\tau)/2} \exp\left(-\frac{1}{2\sigma_0^2}\left(\sum_{i=1}^{\tau}x_i - \mu_0\right)^2 - \frac{1}{2\sigma_1^2}\left(\sum_{i=\tau+1}^{k}x_i - \mu_1\right)^2\right). \tag{10}$$

The MLE of the mean can be computed from the testing data using (Eq. (5)). However, the variance now has to be computed utilizing the MLE for the mean as well [26]:

$$S_{\tau,k}^2 = \frac{1}{k-\tau}\sum_{i=\tau+1}^{k}\left(x_i - \hat{\mu}_{1,\tau,k}\right)^2. \tag{11}$$

As previously stated, from a quality control standpoint only an increase in the variance is of concern, and the MLE for the variance can be computed as follows [26]:

$$\hat{\sigma}^2_{1,\tau,k} = \max\{\sigma_0^2, S^2_{\tau,k}\}. \tag{12}$$

If there are no shifts in the mean for testing data, the variance is computed as follows [26]:

$$S^2_{0,\tau,k} = \frac{1}{k-\tau}\sum_{i=\tau+1}^{k}(x_i - \mu_0)^2. \tag{13}$$

In this case, the GLR statistic designed to simultaneously monitor both shifts in the mean and variance, and can be computed by taking the log-likelihood ratio of (Eqs. (3) and (10)) resulting in the following equation [26]:

$$R_k = \max_{0 \le \tau < k} \frac{k-\tau}{2}\left[\frac{S^2_{0,\tau,k}}{\sigma_0^2} - \frac{S^2_{\tau,k}}{\hat{\sigma}^2_{1,\tau,k}} - \ln\left(\frac{\hat{\sigma}^2_{1,\tau,k}}{\sigma_0^2}\right)\right] \tag{14}$$

It is important to note that for this particular GLR method, two parameters, that is, the mean and the variance have to be estimated using their MLE, since the type of shift is unknown.

### *3.1.4 Multivariate GLR chart for a shift in the mean*

Since using a univariate GLR chart may not always be practical, Wang and Reynolds [27] introduce the multivariate GLR chart, designed to specifically monitor shifts in the process mean for multivariate applications. In this case, the GLR statistic is defined as follows:

$$R_k = \max_{\max(0,k-m) \le t < k}\left(\frac{k-t}{2}(\hat{\mu}_{1,t,k} - \mu_0)\cdot \sum\nolimits_0^{-1}\cdot(\hat{\mu}_{1,t,k} - \mu_0)\right) \tag{15}$$

Where $\mu_0$ is the multivariate mean vector of the process under normal operating conditions, $\hat{\mu}_{1,t,k}$ is the MLE of a sustained process mean shift $\mu_1$ at time index $k$ over sample window of maximum length $m$, and $\sum_0$ is the process covariance matrix under normal conditions [27].

## 3.2 Fault detection using PCA-based GLR

The PCA method introduced in Section 2 is commonly utilized by many industries. Therefore, it is necessary to integrate the simplicity of the PCA method with the advantages brought forward by the GLR charts, so that it can be easily applied to

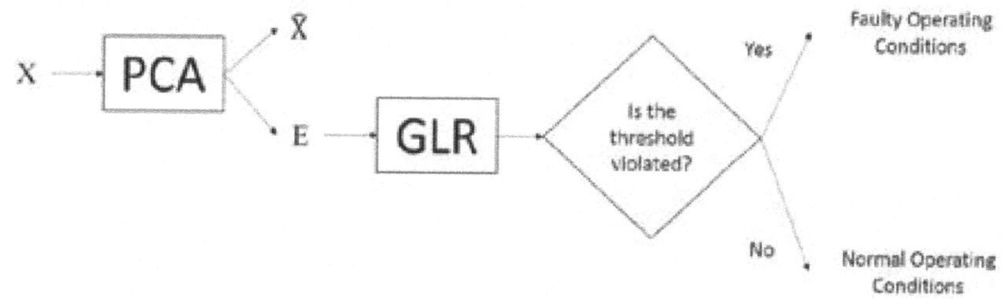


**Figure 1.**
*PCA-based GLR fault detection algorithm.*

monitor processes online. **Figure 1** illustrates the fault detection algorithm utilized in this work.

PCA is utilized in order to model available data. The different GLR charts can then be applied on the residuals produced by the PCA model in order to determine if the process is operating under normal or faulty conditions. The fault detection threshold limits are obtained from an empirical distribution of the GLR statistic computed under normal operating conditions. The residual space is typically better able at detecting faults of smaller magnitude [10].

## 4. Moving window interval data aggregation

Data utilized in the construction of a PCA model may be of two types depending on the application being monitored: single-valued, and interval-valued. Single-valued data can be directly obtained from sensors measuring particular variables in a process, while interval-valued data is aggregated or artificially generated from batch single-valued measurements, thereby resulting in a range of possible measurement values for a given process variable at one time instant [18].

An interval is defined using a lower and upper bound, such as $[a, b]$, where $a \leq b$. In this work, interval data is generated by aggregating the single-valued samples in a dataset, such that the mean of each block of aggregated samples is defined as the interval center ($c$), and the standard deviation of each block of aggregated samples is defined as the interval radii ($r$). Consequently, the intervals can now be defined as $[c - r, c + r]$. Unlike the lower and upper bounds, the centers and radii are of particular importance because they can be used to represent unique characteristics of the classical samples from which they are generated [19].

Initially, the use of interval data is motivated by the need to quickly and efficiently monitor large datasets [28], in addition to its ability to deal with missing values without the need to remove entire samples. Generating intervals by aggregation is a form of batch processing, which may not always be ideal. The ability to monitor faults in real-time is typically much more desirable from a quality and safety standpoint. It also becomes impractical to use batch aggregation when discussing processes with a low sample size or low sampling frequency.

As a result, interval data aggregation must be adapted for real-time monitoring purposes. One way to do that would be to use a moving window aggregation technique, such that any observed sample is aggregated with previously gathered samples, if any, in the defined window size. This allows for the generation and processing of interval data in real-time, without the need to wait for multiple samples to be observed before processing.

As expected, however, this method suffers from some drawbacks relative to its batch aggregation counterpart. The moving window approach may cause smearing along the detection statistic, leading to higher false alarms and lower detection rates. This is especially true for large window sizes, as is the case for most methods which apply that approach. The problem can be mitigated by limiting the window size to reasonable limits, whilst also adjusting the threshold in order to meet the desired false alarm rates of the process.

### 4.1 Integration with PCA-based GLR

Interval principal component analysis (IPCA) methods are an extension to the classical PCA method, and they have been explored in literature for fault detection and isolation examples [29, 30]. In this work, three IPCA methods will be briefly introduced, before discussing our proposed method of integrating the moving window interval approach to the PCA-based GLR technique.

Centers IPCA (CIPCA) was introduced by Cazes et al. [31], where the idea was to only apply PCA to the matrix of interval centers. This method focuses on the variation between the intervals of a dataset, rather than the variations within them [18, 32]. Midpoint-Radii IPCA (MRIPCA) was developed by Lauro et al. [33–36], where PCA models are separately generated for the centers and radii matrices of the interval training dataset. Finally, the Symbolic Covariance IPCA (SCIPCA) method was introduced by Le-Rademacher et al. [18, 32] as a way to better represent the range and variability found in interval data.

In this paper, the integration of the moving window aggregation to PCA-based GLR will be as follows. After generating an interval sample for each single-valued sample, the single-valued matrices of interval centers and radii are extracted. The matrices are then concatenated along the variables dimension, so as to maintain the number of samples, but double the number of variables. This is similar to the MRIPCA method, except it avoids the need to apply PCA twice, eliminating any additional processing complexity.

## 5. Illustrative examples

This section evaluates the performance of the three PCA-based GLR charts described in Section 3, and the moving window aggregation method discussed in Section 4. The PCA-based GLR charts are evaluated under different fault scenarios, and this is done through two illustrative examples: a simulated synthetic data set, and the benchmark Tennessee Eastman Process (TEP). Three fault detection metrics are used to evaluate the performance of each univariate chart: missed DR (which is equal to 100-DR), FAR, and average out-of-control run length (ARL1). Finally, the moving window interval aggregation method, in tandem with the PCA-based multivariate GLR chart, are analyzed using the benchmark TEP process, and the results are tabulated and compared to the single-valued multivariate GLR chart.

### 5.1 Simulated synthetic data example

The purpose of this example is to utilize a simple linear model to compare and evaluate the performance of the difference PCA-based univariate GLR charts. The linear data set can be generated using the following model [37]:

$$\begin{bmatrix} x_1 \\ x_2 \\ x_3 \\ x_4 \\ x_5 \\ x_6 \end{bmatrix} = \begin{bmatrix} -0.3441 & 0.4815 & 0.6637 \\ -0.2313 & -0.5936 & 0.3545 \\ -0.5060 & 0.2495 & 0.0739 \\ -0.5552 & -0.2405 & -0.1123 \\ -0.3371 & 0.3822 & -0.6115 \\ -0.3877 & -0.3868 & -0.2045 \end{bmatrix} \begin{bmatrix} t_1 \\ t_2 \\ t_3 \end{bmatrix} + noise \tag{16}$$

where, $t_1$, $t_2$, and $t_3$, are uniformly distributed random variables with ranges, $[0, 2]$, $[0, 1.6]$, and $[0, 1.2]$, respectively, while the noise follows a normal distribution with zero-mean and standard deviation of 0.2 [37].

The linear model is used to generate 6000 observations, split into training and testing data sets of 3000 observations each. The training data are used to train the PCA model, while the testing data are used to evaluate the performance of all techniques using three cases of faults: a shift in the mean, a shift in the variance, and a simultaneous shift in both.

Five charts are evaluated and compared: the PCA-based $T^2$ and Q charts, and the three different PCA-based univariate GLR charts. The faulty region is highlighted in light blue for all figures, and the fault detection threshold limits for all charts are represented by the red dotted line. For each case a Monte-Carlo simulation of 1000 realizations is carried out in order to obtain meaningful results, so that conclusions can be drawn.

### 5.1.1 *Case 1: a shift in the mean*

For this case, a shift in the mean of $1\sigma$ was introduced between observations 1501 and 3000 in $x_1$ in the testing data set. This fault size was chosen as most conventional techniques are unable to detect a fault of this magnitude. Faults of higher magnitude would likely provide misleading results and exaggerate the robustness of the method in question, leading to a biased comparison.

As can be seen through **Figure 2**, the $T^2$ and Q charts are unable to detect the entirety of the fault. In contrast, two GLR charts (**Figure 3a** and **c**), are able to detect most of the fault, while the GLR chart designed to monitor a shift in the variance (**Figure 3b**) could not detect that a shift in the mean was present.

Examining the summary of the fault detection results (**Table 1**), it can be observed that the GLR chart designed to monitor shifts in the mean (**Figure 3a**) provided the lowest missed DR and $ARL_1$ values, compared to all other charts.

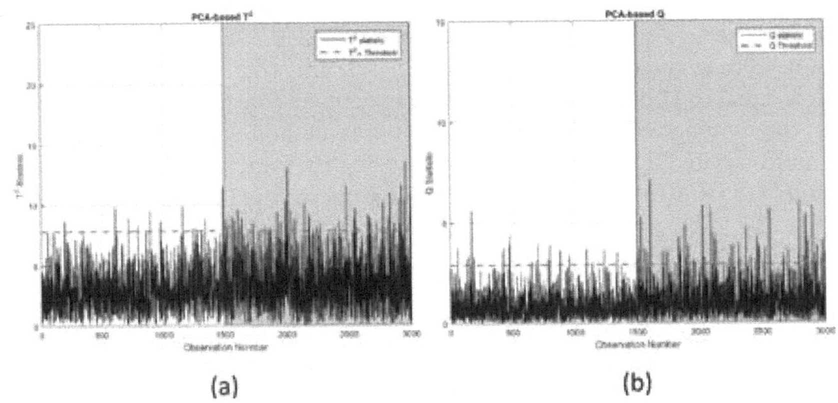


**Figure 2.**
*PCA-based $T^2$ and Q charts (case 1).*

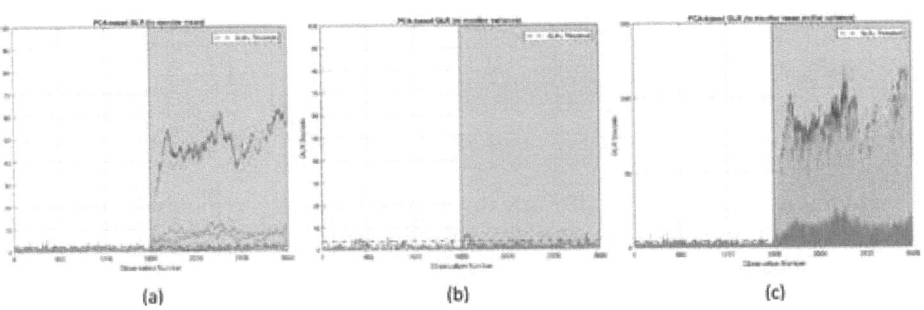


**Figure 3.**
*PCA-based GLR charts (case 1).*

The relatively high missed DR of the GLR chart designed to simultaneously monitor shifts in both the mean and variance (**Figure 3c**) can be attributed to the fact that two parameters need to be estimated from available data while maximizing the GLR statistic, thereby making it difficult to predict a shift in a single parameter as efficiently.

### *5.1.2 Case 2: a shift in the variance*

For this case, an increase in the variance (double that of the training data) was introduced between observations 1501:3000 in $x_1$ in the testing data set. This shift in the variance is too small for detection by most conventional techniques.

As can be seen through **Figure 4**, the $T^2$ and Q charts are unable to detect the entirety of the fault. In contrast, two GLR charts (**Figure 5b** and **c**) were able to detect most of the fault, while the GLR chart designed to monitor a shift in the mean (**Figure 5a**) could not detect it as well. Examining the summary of the results (**Table 2**), it can be observed that the GLR chart designed to monitor a shift in the variance (**Figure 5b**) provided the lowest missed DR and $ARL_1$ values, compared to other charts.

### *5.1.3 Case 3: a shift in the mean and/or variance*

For this case, a simultaneous shift in the mean of $1\sigma$ and an increase in the variance (double that of the training data) was introduced between observations 1501:3000 in $x_1$ in the testing data set.

| | PCA-based $T^2$ | PCA-based Q | PCA-based GLR (to monitor mean) | PCA-based GLR (to monitor variance) | PCA-based GLR (to monitor mean and/or variance) |
|---|---|---|---|---|---|
| Missed DR (%) | 95.3 | 94.5 | 00.4 | 85.1 | 31.5 |
| FAR (%) | 05.2 | 05.5 | 05.3 | 05.8 | 04.6 |
| $ARL_1$ | 20.1 | 16.6 | 04.8 | 81.8 | 05.0 |

**Table 1.**
*Summary of fault detection results (case 1).*

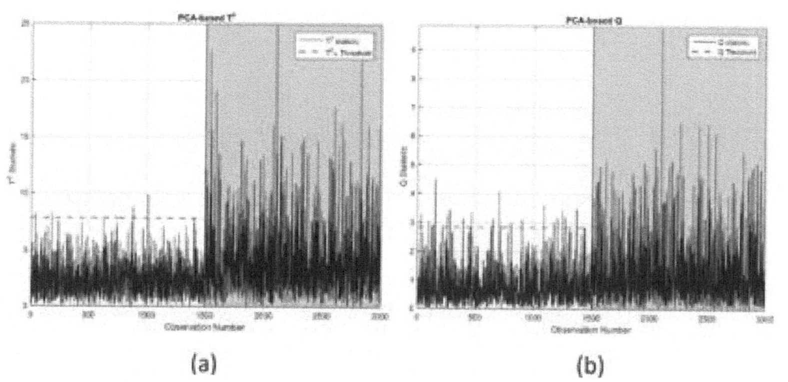

**Figure 4.**
*PCA-based $T^2$ and Q charts (case 2).*

As can be seen through **Figure 6**, the $T^2$ and Q charts are unable to detect the entirety of the fault once more. Although it might seem that all three GLR charts (**Figure 7**) are able to detect most of the fault, upon closer inspection of the results summarized in **Table 3**, it can be observed that the GLR charts designed to independently detect a shift in the mean (**Figure 7a**), and variance (**Figure 7b**), are able to provide significantly lower missed DR and $ARL_1$ values compared to the chart designed to monitors shifts in both (**Figure 7c**).

The main conclusion from this example is that if a process is expected to experience shifts in both the mean and/or variance, it is more beneficial to run the PCA-based GLR charts designed to independently monitor shifts in the mean and variance as two parallel charts, rather than utilizing the GLR chart designed to simultaneously monitor both. Based on this conclusion, only the former two GLR charts will be utilized for the next example.

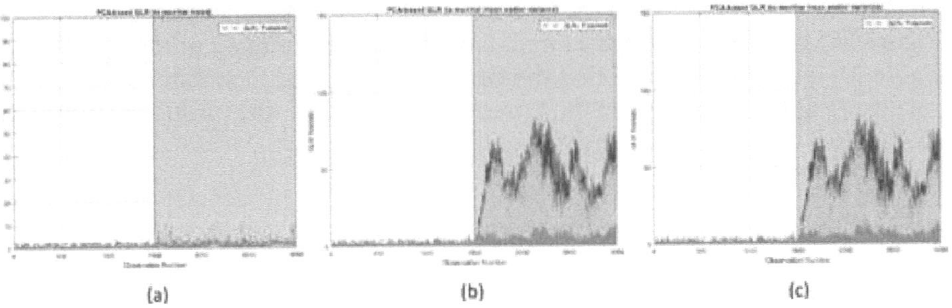


**Figure 5.**
*PCA-based GLR charts (case 2).*

| | PCA-based $T^2$ | PCA-based Q | PCA-based GLR (to monitor mean) | PCA-based GLR (to monitor variance) | PCA-based GLR (to monitor mean and/or variance) |
|---|---|---|---|---|---|
| Missed DR (%) | 90.2 | 88.6 | 47.5 | 00.7 | 33.0 |
| FAR (%) | 05.3 | 05.4 | 05.0 | 04.8 | 04.8 |
| $ARL_1$ | 10.1 | 8.3 | 07.9 | 04.5 | 05.6 |

**Table 2.**
*Summary of fault detection results (case 2).*

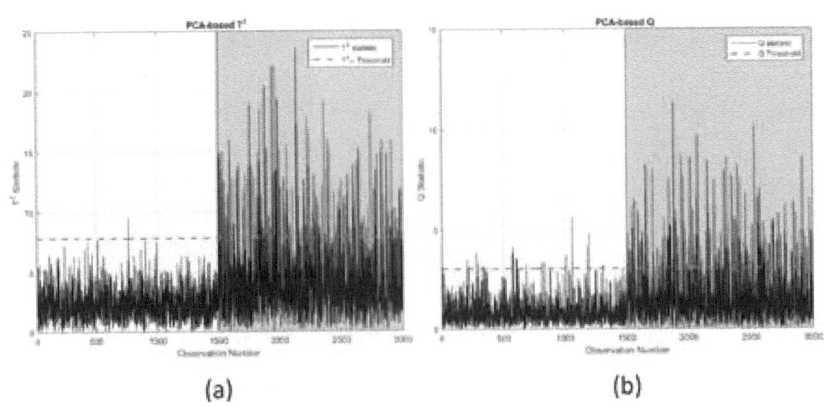


**Figure 6.**
*PCA-based $T^2$ and Q charts (case 3).*

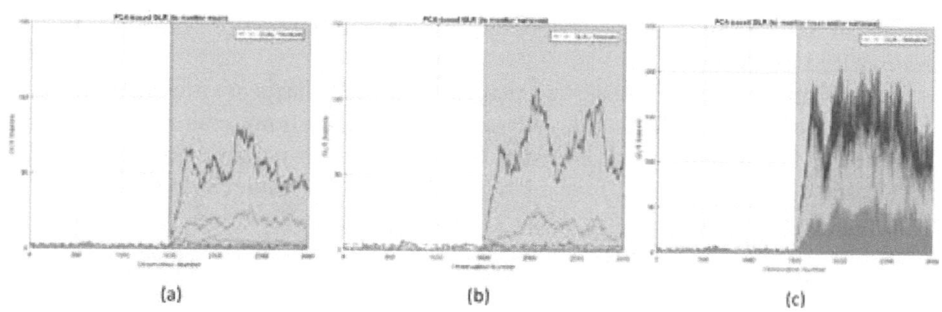

**Figure 7.**
*PCA-based GLR charts (case 3).*

| | PCA-based $T^2$ | PCA-based Q | PCA-based GLR (to monitor mean) | PCA-based GLR (to monitor variance) | PCA-based GLR (to monitor mean and/or variance) |
|---|---|---|---|---|---|
| Missed DR (%) | 86.7 | 84.5 | 00.4 | 00.4 | 24.2 |
| FAR (%) | 05.2 | 05.2 | 04.9 | 05.3 | 05.5 |
| $ARL_1$ | 07.5 | 06.0 | 03.2 | 03.9 | 04.9 |

**Table 3.**
*Summary of fault detection results (case 3).*

## 5.2 Tennessee Eastman Process (TEP)

In order to assess the feasibility of using two separate GLR charts to monitor shifts in the process mean and variance, their performance has to be evaluated using real data. Many authors utilize the Tennessee Eastman Process (TEP) in order to evaluate the performance of their techniques [17, 38, 39]. The Tennessee Eastman Process is a realistic simulation of an actual chemical process that consists of a reactor, condenser, stripper, compressor, and separator, and is widely accepted as a benchmark for fault detection [17].

The Tennessee Eastman Process contains a bank of pre-defined faults that can be utilized by authors in order to assess the performance of their developed fault detection algorithms. More information on the Tennessee Eastman Process, the process description, and the available bank of faults is available in literature [10, 17, 21, 38, 39].

Two fault scenarios will be examined in this work: IDV 3 and IDV 11 [39]. IDV 3 is a shift in the mean of the temperature of Feed D, while IDV 11 is random variation in the reactor cooling water inlet temperature [39]. These two fault scenarios were selected because the conventional techniques are unable to provide the best possible detection. For both scenarios, the fault is introduced after 800 observations of normal operation. The performance of four charts are evaluated: PCA-based $T^2$ and Q charts, and the PCA-based univariate GLR charts designed to independently monitor shifts in the mean and variance. The faulty region is highlighted in light blue in all figures.

### *5.2.1 IDV 3: a step fault in the mean of the temperature of feed D*

For the case where there is a shift in the mean of the temperature of Feed D, the PCA-based $T^2$ and Q charts, and the PCA-based univariate GLR charts are illustrated in **Figures 8** and **9** respectively, and the fault detection results are summarized in **Table 4**.

From **Figure 8** it can be observed that the $T^2$ and Q charts are unable to detect the entirety of the fault, while the GLR chart designed to monitor shifts in the mean (**Figure 9a**) is able to detect the most of the fault, and provides the lowest missed DR (**Table 4**). Although, the $T^2$ chart returns a low $ARL_1$ value, it does not detect the fault efficiently, and the low $ARL_1$ value can be attributed to random noise.

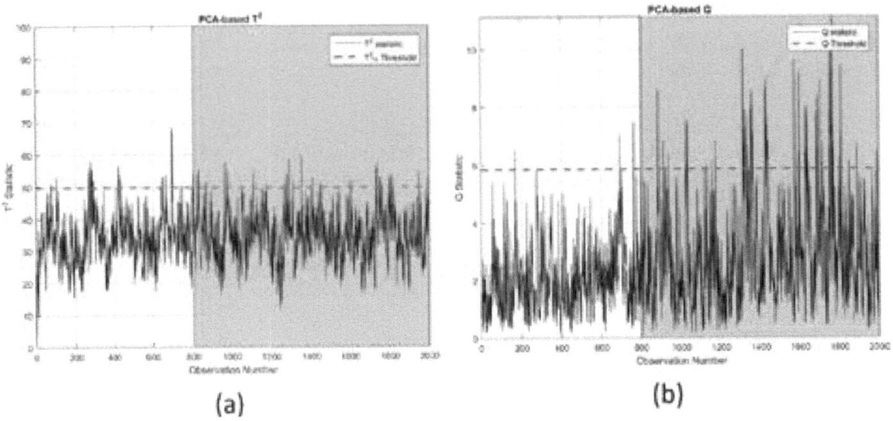

**Figure 8.**
*PCA-based $T^2$ and Q charts (IDV 3).*

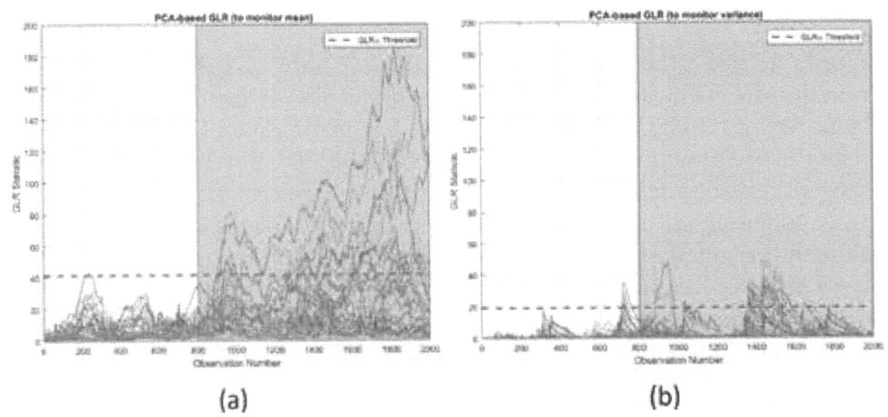

**Figure 9.**
*PCA-based GLR charts (IDV 3).*

| | PCA-based $T^2$ | PCA-based Q | PCA-based GLR (to monitor mean) | PCA-based GLR (to monitor variance) |
|---|---|---|---|---|
| Missed DR (%) | 97.6 | 92.8 | 07.9 | 70.9 |
| FAR (%) | 04.8 | 04.5 | 05.0 | 05.4 |
| $ARL_1$ | 02.0 | 86.0 | 84.0 | 84.00 |

**Table 4.**
*Summary of fault detection results (IDV 3).*

### *5.2.2 IDV 11: random variation in the reactor cooling water inlet temperature*

For the case where there is random variation in the reactor cooling water inlet temperature, the $T^2$ and Q charts, and the GLR charts are illustrated in **Figures 10** and **11** respectively, and the fault detection results are summarized in **Table 5**.

Although it might seem like the $T^2$ and Q charts (**Figure 10**) are able to detect most of the fault, they still have higher missed DR than both GLR charts (**Figure 11**). The GLR chart designed to monitor shifts in the variance provides the lowest missed DR from the charts that were compared.

From this example we can conclude that the PCA-based GLR charts are able to provide improved fault detection results over the conventional PCA-based

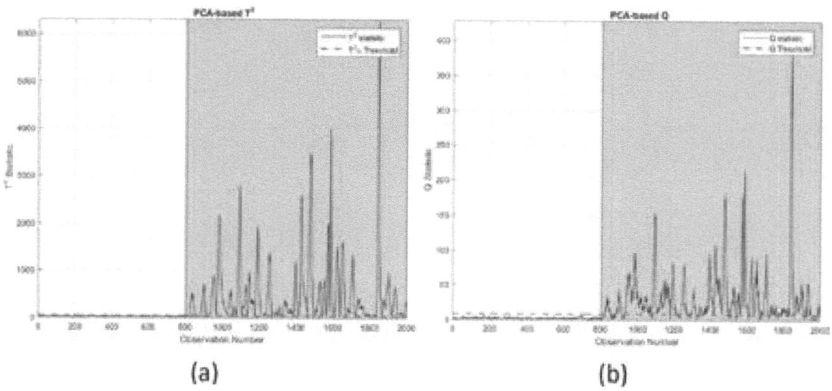


**Figure 10.**
*PCA-based $T^2$ and Q charts (IDV 11).*

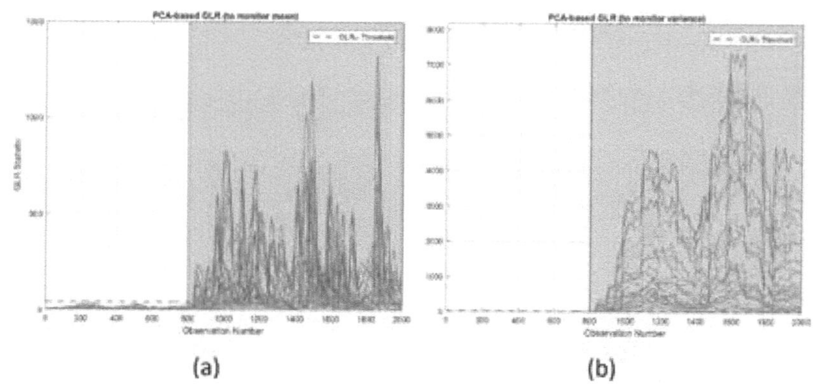


**Figure 11.**
*PCA-based GLR charts (IDV 11).*

| | PCA-based $T^2$ | PCA-based Q | PCA-based GLR (to monitor mean) | PCA-based GLR (to monitor variance) |
|---|---|---|---|---|
| Missed DR (%) | 09.9 | 22.3 | 02.3 | 01.9 |
| FAR (%) | 05.1 | 05.0 | 05.0 | 05.4 |
| $ARL_1$ | 20.0 | 24.0 | 28.0 | 24.0 |

**Table 5.**
*Summary of fault detection results (IDV 11).*

$T^2$ and Q charts. The improved results can be attributed to the use of MLEs to estimate the values of the unknown parameters used to maximize the GLR statistic, allowing for the best possible DR to be achieved for a fixed FAR. This example also demonstrates that the GLR charts can be easily designed and utilized to monitor chemical processes, such as the TEP.

### *5.2.3 IDV 3 and IDV 11: single-valued vs. interval-valued multivariate GLR chart*

For the final case study, the moving window interval aggregation method is tested for the same fault scenarios tested previously for the TEP: IDV 3 and IDV 11. A smaller sample window size of 10 samples is used for the multivariate GLR chart

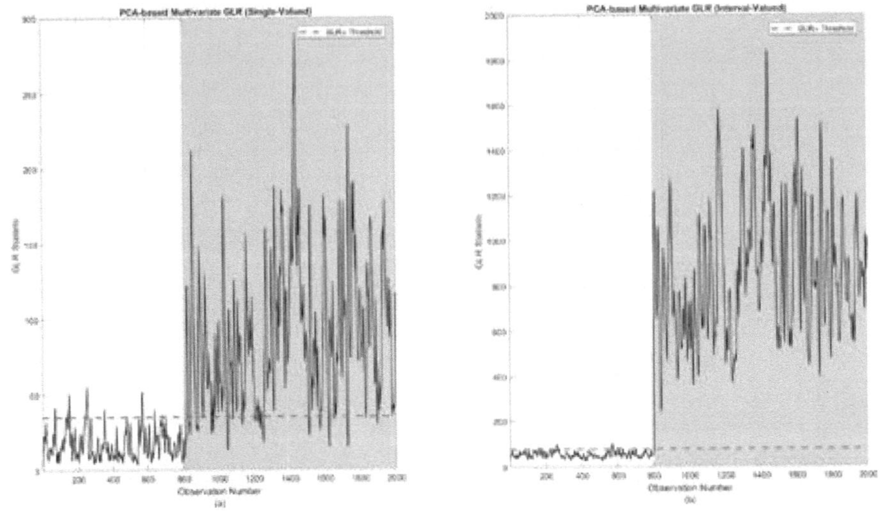


**Figure 12.**
*PCA-based multivariate GLR charts (IDV 3).*

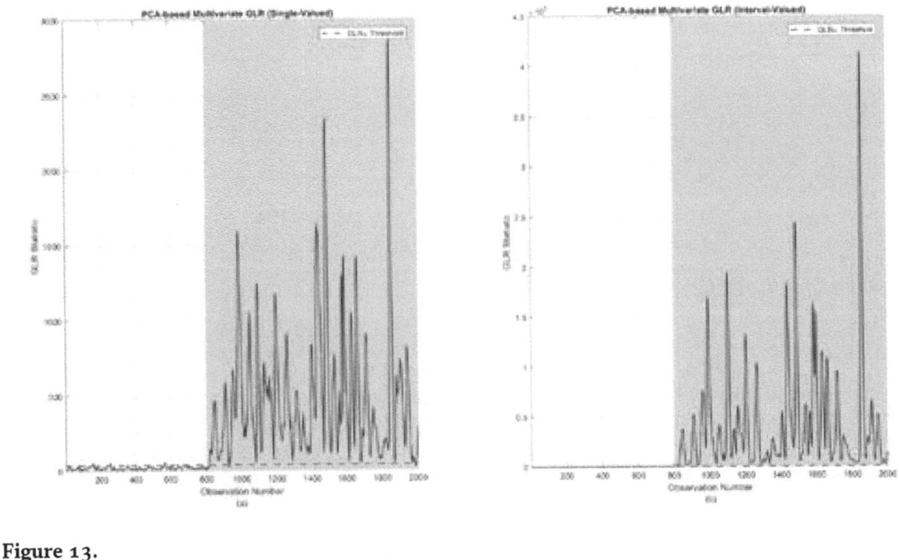


**Figure 13.**
*PCA-based multivariate GLR charts (IDV 11).*

| | IDV 3 Single-valued multivariate GLR | IDV 3 Interval-valued multivariate GLR | IDV 11 Single-valued multivariate GLR | IDV 11 Interval-valued multivariate GLR |
|---|---|---|---|---|
| Missed DR (%) | 15.1 | 00.0 | 02.0 | 00.0 |
| FAR (%) | 05.0 | 05.0 | 05.0 | 05.0 |

**Table 6.**
*Summary of fault detection results (single vs. interval data) for* $\alpha = 5\%$.

in order to highlight the difference between using single and interval-valued data more clearly.

The interval aggregation window size was set at 10 samples. The IDV 3 and IDV 11 scenarios for both data types are shown in **Figures 12** and **13**, and the metrics for each method are tabulated in **Table 6**.

There are two major observations to be made from the results. First, the use of the multivariate GLR chart allowed for a more stable FAR for all cases due to the presence of a single statistic to monitor for all variables, as opposed to the one for each variable when using the univariate GLR charts. Second, the missed DR when using interval data was significantly lower than that for single-valued data, reaching perfect performance levels of zero missed DR for both scenarios.

The latter observation is attributed to interval data, especially the method of generation, where the centers and radii are used as independent variables in the same dataset. This method of aggregation helps the PCA model account for shifts in the mean and variance respectively, similar to the univariate GLR chart outline in Section 3.1.3. However, it does so without the need to tune any extra parameters, due to the fact that a fault in the centers is likely to be caused by a shift in the mean, while a fault in the radii is likely to be caused by a shift in the variance.

## 6. Conclusions

In this chapter, the performance of GLR charts were compared to conventional fault detection statistics, specifically the Q and $T^2$ statistics, and the integration of interval-valued data into real-time process monitoring was explored. The performance of different PCA-based univariate GLR charts were examined using single-valued data through two illustrative examples: simulated synthetic data, and the Tennessee Eastman Process. The performance of the moving window interval aggregation method was evaluated alongside that of single-valued data for the multivariate GLR chart as well.

The results demonstrate that in order to monitor processes that may experience both shifts in the mean and/or variance, the best performance is achieved by implementing the two respective univariate GLR charts separately in parallel, rather than the single chart designed to simultaneously detect shifts in both, as the simultaneous estimation of two parameters is unable to provide the best possible fault detection performance. Moreover, the moving window interval aggregation method, when combined with the multivariate GLR chart, was able to provide a perfectly stable statistic, with an unwavering false alarm rate, in addition to the best possible performance in detecting shifts in the mean and variance for two scenarios of the Tennessee Eastman Process.

## References

[1] Montgomery DC. Introduction to Statistical Quality Control. 7th ed. Hoboken, NJ: John Wiley and Sons; 2013

[2] Chakrabarty A, Mannan S, Cagin T. Multiscale Modeling for Process Safety Applications. 1st ed. Oxford, United Kingdom: Butterworth-Heinemann; 2015

[3] Venkatasubramanian V, Rengaswamy R, Yin K, Kavuri SN. A review of process fault detection and diagnosis: Part I: Quantitative model-based methods. Computers and Chemical Engineering. 2003;**27**:293-311

[4] Venkatasubramanian V, Rengaswamy R, Yin K, Kavuri SN. A review of process fault detection and diagnosis: Part II: Qualitative models and search strategies. Computers and Chemical Engineering. 2003;**27**:313-326

[5] Venkatasubramanian V, Rengaswamy R, Yin K, Kavuri SN. A review of process fault detection and diagnosis: Part III: Process history based methods. Computers and Chemical Engineering. 2003;**27**:327-346

[6] George JP, Chen Z, Shaw P. Fault detection of drinking water treatment process using PCA and Hotelling's T2 chart. International Journal of Computer and Information Engineering. 2009;**3**: 970-975

[7] Sanguansat P, editor. Principal Component Analysis: Multidisciplinary Applications. Rijeka: InTech; 2009. DOI: 10.5772/2694

[8] Sanguansat P, editor. Principal Component Analysis: Engineering Applications. Rijeka: InTech; 2012. DOI: 10.5772/2693

[9] Joliffe IT. Principal Component Analysis. 2nd ed. New York, NY: Springer-Verlag; 2002

[10] Sheriff MZ, Mansouri M, Karim MN, Nounou H, Nounou M. Fault detection using multiscale PCA-based moving window GLRT. Journal of Process Control. 2017;**54**:47-64. DOI: 10.1016/j.jprocont.2017.03.004

[11] Sheriff MZ, Botre C, Mansouri M, Nounou H, Nounou M, Karim MN. Process monitoring using data-based fault detection techniques: Comparative studies. In: Fault Diagnosis Detect. InTech; 2017. DOI: 10.5772/67347

[12] Mansouri M, Sheriff MZ, Baklouti R, Nounou M, Nounou H, Ben Hamida A, et al. Statistical fault detection of chemical process: Comparative studies. Journal of Chemical Engineering and Process Technology. 2016;**07**:1-10. DOI: 10.4172/2157-7048.1000282

[13] Botre C, Mansouri M, Nounou M, Nounou H, Karim MN. Kernel PLS-based GLRT method for fault detection of chemical processes. Journal of Loss Prevention in the Process Industries. 2016;**43**:212-224. DOI: 10.1016/j.jlp.2016.05.023

[14] Tharrault Y, Mourot G, Ragot J. Fault detection and isolation with robust principal component analysis. In: 2008 16th Mediterr. Conf. Control Autom. Vol. 18. 2008. pp. 429-442. DOI: 10.1109/MED.2008.4602224

[15] Montgomery DC, Runger GC. Applied Statistics and Probability for Engineers. 5th ed. Hoboken, NJ: John Wiley and Sons, Inc.; 2011

[16] Harrou F, Nounou MN, Nounou HN. Detecting abnormal ozone levels using PCA-based GLR hypothesis testing. In: Proc. 2013 IEEE Symp. Comput. Intell. Data Mining, CIDM 2013–2013 IEEE Symp. Ser. Comput. Intell. SSCI 2013; 2013. pp. 95-102. DOI: 10.1109/CIDM.2013.6597223

[17] Downs JJ, Vogel EF. A plant-wide industrial process control problem. Computers and Chemical Engineering. 1993;**17**:245-255. DOI: 10.1016/0098-1354(93)80018-I

[18] Le-Rademacher JG. Principal Component Analysis for Interval-Valued and Histogram-Valued Data and Likelihood Functions and some Maximum Likelihood Estimators for Symbolic Data. Athens, GA: University of Georgia; 2008

[19] Basha N. Interval Principal Component Analysis and its Application to Fault Detection and Data Classification. College Station, TX: Texas A&M University; 2018

[20] Strang G. Introduction to Linear Algebra. 5th ed. Wellesley, MA: Wellesley-Cambridge Press; 2016

[21] Russell EL, Chiang LH, Braatz RD. Fault Detection and Diagnosis in Industrial Systems. New York, NY: Springer-Verlag; 2001

[22] Hotelling H. Analysis of a complex of statistical variables into principal components. Journal of Education & Psychology. 1933;**24**:417-441. DOI: 10.1037/h0071325

[23] Harrou F, Nounou MN, Nounou HN, Madakyaru M. Statistical fault detection using PCA-based GLR hypothesis testing. Journal of Loss Prevention in the Process Industries. 2013;**26**:129-139. DOI: 10.1016/j.jlp.2012.10.003

[24] Reynolds MR, Lou JY. An evaluation of a GLR control chart for monitoring the process mean. Journal of Quality Technology. 2010;**42**:287-310

[25] Reynolds Jr MR, Lou J. A GLR control chart for monitoring the process variance. In: Lenz HJ, Schmid W, Wilrich, editors. Frontiers in Statistical Quality Control. New York, NY: Springer; 2012;**10**:3-17. DOI: 10.1007/978-3-7908-2846-7

[26] Reynolds MR, Lou J, Lee J, Wang S. The design of GLR control charts for monitoring the process mean and variance. Journal of Quality Technology. 2013;**45**:34-60

[27] Wang S, Reynolds MR. A GLR control chart for monitoring the mean vector of a multivariate normal process. Journal of Quality Technology. 2013;**45**:18-33

[28] Billard L, Le-Rademacher J. Principal component analysis for interval data. Wiley Interdisciplinary Reviews: Computational Statistics. 2012;**4**:535-540. DOI: 10.1002/wics.1231

[29] Benaicha A, Guerfel M, Bougila K, Benothman N. New PCA-based methodology for sensor fault detection and localization. In: 8th Int. Conf. Model. Simul.; Hammamet: Tunisia; 2010

[30] Izem TA, Bougheloum W, Harkat MF, Djeghaba M. Fault detection and isolation using interval principal component analysis methods. IFAC-PapersOnLine. 2015;**48**:1402-1407. DOI: 10.1016/j.ifacol.2015.09.721

[31] Cazes P, Chouakria A, Diday E, Schektman Y. Extension de l'analyse en composantes principales à des données de type intervalle. Revue de Statistique Appliquée. 1997;**45**:5-24

[32] Le-Rademacher J, Billard L. Symbolic covariance principal component analysis and visualization for interval-valued data. Journal of Computational and Graphical Statistics. 2012;**21**:413-432. DOI: 10.1080/10618600.2012.679895

[33] Lauro CN, Palumbo F. Principal component analysis for non-precise data. In: New Developments in

Classification and Data Analysis. Berlin/Heidelberg: Springer-Verlag; n.d. pp. 173-184. DOI: 10.1007/3-540-27373-5_21

[34] Lauro CN, Palumbo F. Principal component analysis of interval data: A symbolic data analysis approach. Computational Statistics. 2000;**15**:73-87. DOI: 10.1007/s001800050038

[35] Lauro NC, Verde R, Irpino A. Principal component analysis of symbolic data described by intervals. In: Symbolic Data Analysis and the SODAS Software. Chichester, UK: John Wiley and Sons, Ltd; n.d. pp. 279-311. DOI: 10.1002/9780470723562.ch15

[36] Palumbo F, Lauro CN. A PCA for interval-valued data based on midpoints and radii. In: New Developments in Psychometrics. Japan, Tokyo: Springer; 2003. pp. 641-648. DOI: 10.1007/978-4-431-66996-8_74

[37] Alcala CF, Joe Qin S. Analysis and generalization of fault diagnosis methods for process monitoring. Journal of Process Control. 2011;**21**:322-330. DOI: 10.1016/j.jprocont.2010.10.005

[38] Lyman PR, Georgakis C. Plant-wide control of the Tennessee Eastman problem. Computers and Chemical Engineering. 1995;**19**:321-331. DOI: 10.1016/0098-1354(94)00057-U

[39] Yin S, Ding SX, Haghani A, Hao H, Zhang P. A comparison study of basic data-driven fault diagnosis and process monitoring methods on the benchmark Tennessee Eastman process. Journal of Process Control. 2012;**22**:1567-1581. DOI: 10.1016/j.jprocont.2012.06.009

Chapter 7

# Real-Time Fault Detection and Diagnosis Using Intelligent Monitoring and Supervision Systems


## Abstract

In monitoring and supervision schemes, fault detection and diagnosis characterize high efficiency and quality production systems. To achieve such properties, these structures are based on techniques that allow detection and diagnosis of failures in real time. Detection signals faults and diagnostics provide the root cause and location. Fault detection is based on signal and process mathematical models, while fault diagnosis is focused on systems theory and process modeling. Monitoring and supervision complement each other in fault management, thus enabling normal and continuous operation. Its application avoids stopping productive processes by early detection of failures and by applying real-time actions to eliminate them, such as predictive and proactive maintenance based on process conditions. The integration of all these methodologies enables intelligent monitoring and supervision systems, enabling real-time fault detection and diagnosis. Their high performance is associated with statistical decision-making techniques, expert systems, artificial neural networks, fuzzy logic and computational procedures, making them efficient and fully autonomous in making decisions in the real-time operation of a production system.



**Keywords:** automatic control, availability, intelligent systems, monitoring, predictive maintenance, supervision


## 1. Introduction

Advances in production techniques have improved the capacity of the productive systems of the industries, since the equipment used in these processes have improved their reliability and availability in the operation, making the productive processes more efficient.

One of the most critical questions about automated system design today is reliability and availability of a system. A traditional way to improve the reliability and availability of systems is to improve the quality, reliability, and robustness of the individual components of such a system, such as such as sensors, actuators, controllers and/or computers, used integrally in modern monitoring processes. Even so, a fault-free operation cannot be guaranteed. Process monitoring and fault diagnosis are a vital part of the innovative and modern systems of automatic management of the operation of production systems [1, 2].

Since the life cycle stages of production process equipment require high investments, and maintenance and operation procedures to achieve appropriate return times on the investments made, must ensure high availability and reliability rates. These performance indexes are improved by reducing the number of failures and managing their severities, while ensuring an increase in overall security.

To achieve these goals, two important techniques are available that allow optimized maintenance management, known as predictive and proactive, which are complemented by the techniques: corrective and preventive. This set of techniques offers its best results through the implementation of efficient real-time monitoring and supervision structures, making production systems highly reliable in supplying their products and in the quality of products offered. Corrective maintenance corrects the problem, preventive maintenance prevents the problem.

On the other hand, predictive maintenance consists in the frequent measurement of physical quantities, considered representative and through the analysis of their behavior, to extract their state or operative condition. This allows to suggest the most appropriate moment to apply the necessary actions in the equipments that present characteristics of being in the initial state of a fault - early failure (the root cause is slightly impacting the equipment continuously), anticipating in this way to the emergence of a serious system failure. The predictive maintenance process allows obtaining a report on the operational condition of the equipment. This process of issuing the report basically comprises four stages:

- Identification of the failure modes that are occurring;
- Fault location;
- Evaluation of its extension;
- Estimation of the remaining life of the equipment or component in question.

In traditional predictive maintenance processes, all these steps are performed manually. Alternatively, these steps can be performed using computer systems that allow automating this process is called Systems for Automatic Fault Diagnosis [3].

As can be inferred, the selection, implementation, operation and maintenance of a System for Automatic Diagnosis of Failures is not a simple task, requiring at each stage, care so that the result provided by the system, after its implementation, is within the one initially specified. For this, it is necessary to use appropriate tools and strategies, in each step, in order to maximize the success in executing each of them.

Proactive maintenance is a procedure that minimizes the impact of lack of maintenance or reduced maintenance on the equipment of a production system and also by its own characteristics complements the other maintenance techniques. The main action of this maintenance is to analyze the performance indicators and identify the root cause of the failures, the degradation of the equipment and to remove them before the severity of a fault itself increases [4].

In this chapter, a description will be given of the various methodologies for converting an online monitoring and supervision system into an intelligent system that allows the detection and diagnosis of failures, training it to assure autonomy in taking the necessary actions in real time to avoid them and seek their causes to eliminate them.

The proposed content has two basic objectives: to discuss some important factors for the success in the implantation and use of these structures or systems, as well as the main benefits in the integrated and simultaneous use of the monitoring and supervision of several physical quantities of the equipment, with the goal of increasing the "accuracy" of fault detection and diagnosis.

The technological development in this area has allowed the emergence of innovative methodologies for the detection and diagnosis of failures. The failure detection method recognizes that the failure has occurred, and fault diagnosis finds the root cause and location of that failure. In general, fault detection methods are based on mathematical models of signal and process, and on methods of systems theory and process modeling to generate fault symptoms. Fault diagnosis methods use causal relationships between fault and symptom, applying statistical decision methods, artificial intelligence and computational software [5].

Among the existing model-based fault diagnosis schemes, the so-called observations-based technique has received much attention since the 1990s. This technique was developed within the framework of the successful theory of advanced control, where powerful tools are available to design or to extrapolate recorded observations through efficient and reliable algorithms for data processing in order to reconstruct process variables.

The content described here is intended to provide an introduction to advanced monitoring and supervision, focused as a framework or intelligent assembly for fault detection and diagnostics [1, 6], and fault-tolerant systems especially for processes characterized by continuous and sampled (discrete) signals.

In general, almost all physical signals are continuous, for example, position and velocity of a body, speech or music picked up by a microphone, voltage or current in an electric circuit.

The sampling (instantaneous) of an analog signal or waveform is the process by which the signal is represented by a discrete set of numbers. These numbers, or samples, are equal to the signal value at well-determined instants (the sampling times). Samples must be obtained in such a way that it is possible to reconstruct the signal accurately. That is, the original waveform, defined in "continuous" time, is represented in "discrete" time by samples obtained at conveniently spaced sampling instants.

An application-oriented approach will also be done with methods that have proven their proper performance in practical applications.

## 2. Monitoring and supervision of systems

The monitoring and supervision of processes aim to show the real state of the equipment involved in a productive process, indicating undesirable or illicit states and the appearance of a change in its initial phase (early failure). This situation will require taking appropriate and immediate action to avoid catastrophic damage in the future.

Deviations from the normal behavior of the parameters of an equipment or system arise from faults and/or errors, which can be attributed to several causes. These changes are symptoms of possible early failure, and if the necessary actions are not taken to eliminate them, they may become actual failures that may compromise the performance of productive systems. The justification for monitoring and supervision systems is to avoid such defects or failures in systems by collecting continuous information (provided by the monitoring system) in real time, on the behavior of the equipment of a production system and its supervision data) that will allow you to determine if a device or equipment is operating normally or at risk.

Deviations from the normal behavior of the parameters of an equipment or system arise from faults and/or errors, which can be attributed to several causes.

These changes are symptoms of possible failures in their early state, and failure to take the necessary actions to eliminate them can lead to real failures that may compromise the performance of productive systems. The justification for the monitoring and supervision systems is to avoid these defects or failures in the systems by collecting continuous information (provided by the monitoring system) in real time, on the behavior of the equipment of a production system and its supervision (data evaluation collected) to determine whether a device or equipment is operating normally or at risk.

The content presented in this chapter is focused primarily on the areas of system monitoring and supervision. We have shown the changes that can be made in these two areas of observation and analysis of the behavior of the parameters of a system during its operation, to make them more efficient in solving problems of production systems. The fundamental objective of this information is to integrate these two areas into one set only through the use of the Smart System technique, allowing its unified application in real time in decision making, in any area of a production system [2].

The Smart System technique will make productive systems economically efficient by improving their performance, quality, reliability of supply, operational flexibility, safety, etc.

## 2.1 Fault diagnosis monitoring systems

It should be noted that the selection, implementation, operation and maintenance of a system for automatic fault diagnosis is a complex task [2, 7]. You must ensure that the result provided by this system is within the programmed specifications. For this, it is necessary to use appropriate tools and strategies, in each step, in order to maximize the success in executing each of them.

The concept of predictive maintenance is directly linked to the monitoring of the condition (state) of one or more equipment. O monitoring as such is a basic tool for the implementation of predictive maintenance strategies. Monitoring can be classified from the point of view of the type of sensor installation (permanent or mobile), or be classified by the data acquisition strategy "continuous/on-line" or "periodic/off-line".

"Continuous/on-line" monitoring systems often work in an integrated way with the Supervisory and Control Systems, or "Supervisory Systems" of the production systems, both of which have individual requirements for data acquisition and functions totally different from one another. The integration of these two systems allows for the "continuous" acquisition of operating data and the variables of slow variation (temperatures, levels, position values, static pressures, etc.) normally available in these systems.

Automatic Diagnostic Systems - ADS are the next step to pure and simple monitoring. These more advanced systems receive information from the monitoring system and, through the use of intelligent software, can manage" Knowledge Bank", where information obtained from various physical parameters is crossed and integrated, from where a result that is closer to what one really wants: an effective aid to decision-making.

## 2.2 Main features of automatic fault diagnostic systems

Automatic processing systems are integrated by computer programs, focused on the technique of artificial intelligence, and are responsible for automatically processing all information from the monitoring systems. The main objective of the integration of these systems in the operation of a productive system is the automatic detection of incipient faults and their main characteristics, that is, faults that are in the initial phase of their formation (early faults), their identification, location and estimation of the degree of severity.

The main characteristic of ADSs is that they can handle large amounts of data generated by Monitoring Systems in a systematic, frequent and automatic way, and optimize the process of data storage during long periods of operation (months or years). Another attribute of the ADSs is their intrinsic characteristic, that is, throughout the time of use, each time less need of the interference of the user. Another important feature of ADSs is their adequacy as a Knowledge Management tool in predictive maintenance [8, 9].

The characteristic limitation of this type of system, as well as of any type of Monitoring System traditionally used, is presented when dealing with faults of instant or catastrophic evolution. For this, "Protection Systems", with fixed and well-established alarm limits, should be considered as the main option. The principles of operation, as well as the necessary technical characteristics, relating to the acquisition, communication and processing of data from each of these systems are fundamentally different and should not be confused [10].

Basically, ADSs have the function of reporting the occurrence of failures when they are still in their infancy, while the Protection Systems must act at the moment an unacceptable operating situation occurs.

The technological development in the systems of monitoring and supervision will allow the structuring and optimized evolution of the areas of automated detection and diagnosis, this being the next step to the pure and simple monitoring. These systems receive information from the monitoring system, consisting essentially of sensors and through the use of the technique of intelligent systems and expert systems, a "knowledge bank" is managed or also called the knowledge base for decision making.

The evaluation of the information provided by the monitoring and supervision system will allow to detect and locate a problem and diagnose its root cause, simultaneously, it will be possible to select the best action to mitigate changes in the behavior of the parameters of interest and eliminate the cause that produces them. Finally, the system itself will decide whether to take this action online or offline, depending on the severity and robustness of the problem.

Another important characteristic that is considered in the design of these systems is their intrinsic characteristic that, throughout the time of use, they are less and less required to interfere with the user. That is, while in the case of traditional monitoring systems, the accumulation of stored, non-processed data by the user is a natural consequence of the monitoring process itself, and the effort to treat such data never diminishes over time. In the ADSs the manual work involved in the processing of information is decreasing over time. This is due to the fact that there are tools and mechanisms of retention and improvement of the knowledge registered in these systems. Thus, by using knowledge management tools, maintenance team members can track, correct, insert, retrieve, and refine existing content in their Knowledge Bank (expert systems).

In this way, it can be said that the joint monitoring and supervision system has become intelligent and consequently autonomous to operate a production system in an efficient way from a technical as well as an economic point of view.

### 2.3 Structuring of monitoring and supervision systems in an intelligent system

In this section, a description of the methodology used for the conversion of a monitoring and supervision system in an intelligent system that allows the detection, localization and diagnosis of failures is made possible to take the most appropriate actions to eliminate them and to seek their causes to avoid them. This will improve the efficiency of production systems.

Smart System or Smart Grid in general terms is the application of information technologies in production systems, integrated with communication systems and with an automated network infrastructure. This technique requires the installation of sensors in all the fundamental equipment of the production systems, structuring a reliable two-way communication system with wide coverage with the various devices and automation of the physical assets.

The current sensors have chips that detect information about the behavior of the parameters of certain equipment. These devices collect the information and those with changes are sent to an operation center through a communication system where they are analyzed to determine what is significant.

This process must occur in real time and online mode and in the presence of significant information, a centralized analysis system (specialized software) will evaluate them and determine the changes that have occurred and what should be done to improve the performance of a given parameter.

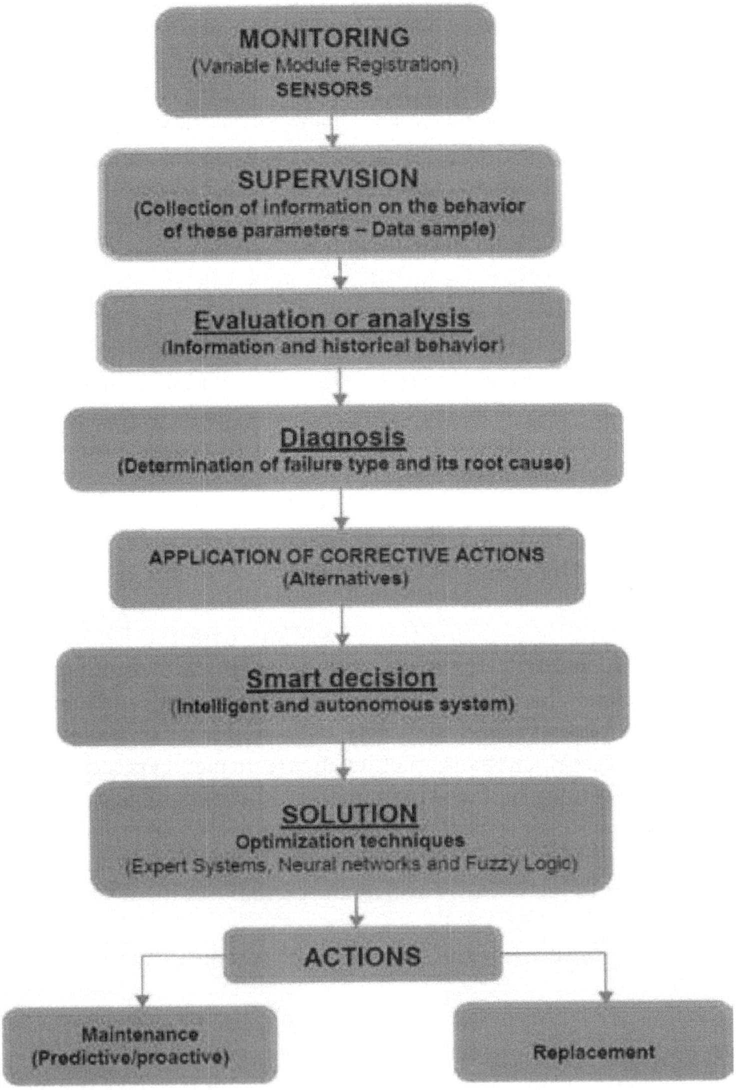


**Figure 1.**
*Flowchart of intelligent monitoring and supervision system.*

In **Figure 1**, a block diagram is presented where the sequential structure of an intelligent monitoring and supervision system is described at a macro level. This configuration is a technological innovation in the area of intelligent automation. Its implementation is done through computational software of reference that will help in the process of evaluation, detection, location, diagnosis and application of the most appropriate actions in the elimination of a problem or failure.

An overview of each of the steps that will make up the intelligent monitoring and supervision systems (see flowchart in **Figure 1**) will be presented, highlighting the methodologies and techniques that will be used in each one of them in order to reach the required efficiency level which allows solving the various problems that arise in the equipment used in the production processes. This efficiency will be measured by the degree of automatism in real time and the autonomy in decision making in the presence of a certain disturbance. This will indicate a fully intelligent system that will safeguard the integrity and security of a production system, avoiding collapses and economic and technical damages.

### *2.3.1 Monitoring*

To monitor is to observe, analyze and be aware of possible signs that something is not normal. In information technology, "not normal" can indicate unavailability of one or more parts of a system or simply change a parameter of a device.

In this phase the observation of changes or changes in the modules of the parameters of transcendence in time is realized. These changes must be recorded within a data collection system called the Database, which will allow us to construct a history of the behavior in time of a given variable or parameter according to a reference level or threshold of behavior [11, 12].

This process is carried out only through a robust system of sensors, installed at strategic points of equipment or system, allowing its observability continuous in time [1].

Monitoring is carried out using the following methodologies:

a. *Digital recorders*—perform digital recording of all information from the sensors. It is through these devices that the history of the behavior of a parameter or variable in time is constructed. This information is usually stored in the binary system.

b. *Remote digital sensors (threshold)*—transducer is the name given to a sensor or actuator, which in turn are devices for detection and actuation in a given process.

With the advent of microcontrollers and microprocessors and the great availability of tools and resources for the processing of digital systems, it was possible to introduce a high computing capacity to the transducers.

The intelligent transducer, which is the integration of: (a) an analog or digital sensor or an actuator, (b) a processing unit, and (c) a network interface.

A smart transducer transforms the sensor's raw signals into a standardized digital representation, transmitting this digital signal to its users through a standardized digital communication protocol.

In **Table 1**, the sensors that are part of the two types of sensors existent for the realization of the monitoring process are described.

c. *Oscillography*—aims to enable the post-event analysis of disturbances, different from the protection systems that must act in real time in response to

| Active sensors | Passive sensors |
|---|---|
| Thermoelectric | Resistive |
| Piezoelectric | Capacitive |
| Pyro electric | Inductive |
| Photovoltaic | Resonant |
| Electromagnetic | |
| Hall effect | |

**Table 1.**
*Types of sensors.*

disturbances. In fact, oscillography is a complementary tool to the protection systems, as it allows the specialist in the analysis of disturbances to verify the adjustments of a given protection, as well as any defects that may arise.

A very useful calculation performed by specialists from the oscillograms is to determine the distance at which a disturbance occurred. In this case, the specialist informs the maintenance team in which region of the transmission line it must act in order to repair the damage caused by the disturbance, making its work easier and more efficient. In addition, the expert performs other procedures, such as the phasor analysis to verify the balance between the phases and the harmonic analysis to observe the intensity of the harmonics present in the signal.

The digitalization of the oscillography signals motivated the growth of the number of computational tools developed to aid in the analysis of perturbations, also allowing the development of sophisticated signal processing tools and intelligent processing systems.

Nowadays the use of oscillography has become quite frequent for the recording of events in production systems (electrical systems, mechanical systems, etc.), since it is possible to observe the development sequence of an event and the interaction between the elements of the system that are part of the event. This implies the progressive growth of the number of oscillography files.

Thus, the need arises to study and develop compression methods with the purpose of reducing the space needed to store these files and make better use of the resources. It is proposed the use of a compression method by synthesis of oscillography files, using redundant adaptive decompositions, which provide a coherent representation with the phenomena present in the recorded signals. These decompositions were based on the technique of Matching Pursuits (MP).

Remote and real time monitoring allows observation of data related to operating conditions, mechanical parameters (fuel, temperature, engine speed, oil pressure and level, vibration, etc.), electrical parameters (currents, powers, voltages, oscillations, etc.), hydraulic parameters (flow, cavitations, water hammer, etc.) and operating hours.

### *2.3.2 Supervision*

It is a process that performs the analysis of collected data for detection of unwanted or non-permitted states. It is searched if a parameter is within the permitted limits or if there are unusual variations.

The supervision area receives information from the Monitoring System and, through the use of intelligent software, a "Knowledge Bank" must be managed, where information obtained from various physical parameters is crossed and integrated, from which a result that is closer to what one really wants: an effective aid to decision-making.

The supervisory system automatically processes the information collected by the monitoring system through internal routines using intelligent techniques (Computational Intelligence). The objective is the automatic detection of incipient faults, that is, early detection of faults, their identification, location and estimation of the degree of severity [4, 13].

Within the area of supervision three very significant and decisive procedures are performed in solving the problems of a productive system:

1. Evaluation of the information collected;

2. Diagnosis of the changes present in these collected data - the system will inform if they are faults in their precocious state or catastrophic failures and their root cause. It will also inform your location;

3. Elimination of failures and their root cause. This will prevent damage and collapse in a production process. Here will be decided the actions that must be taken and depending on the severity of the failure its execution will be online or offline.

*2.3.2.1 Evaluation*

At this stage, the detection of variations or changes in the normal behavior of a parameter is carried out. For this, a comparison process is carried out with a previously defined reference value. This revision is performed in real time and its result compared to the past behavior, this will allow defining if it is really presenting an abnormality or simply it is an isolated eventuality.

The behavior history is analyzed and an image of the state of the selected parameters is created [8, 9]. This image is compared to the behavior of these same parameters in real time. This part of the evaluation uses only stored (digital) data that reaches a threshold value (reference value).

An evaluation of the recorded oscillograms is also performed, interpreting all recorded graphs related to the behavior of a given parameter.

To perform the database evaluation process with the historical behavior of the parameters of interest, there are tools or methodologies that allow the execution of this process in an optimized and efficient manner. These methodologies are as follows and depend greatly on the type of signal being monitored:

a. Statistical and Probabilistic Techniques

At this stage of the research the most used statistical method is least squares. *Applies mainly to processes with linear characteristics*. The probabilistic methods most commonly used to determine the probabilities of states and the probability density function of an equipment or system are: (a) state space method or Markovian process that describes states and possible transitions between them, (b) Monte Carlo simulation performs several computational simulations of a process for a certain period, ending the simulations procedure, estimating the desired indices as the probability of a failure to occur, its frequency and the duration of the failure.

b. Kalman filter

The Kalman filter produces estimates of the actual values of measured quantities and associated values, predicting a value, estimating the uncertainty of the predicted value, and calculating a weighted average between the predicted value and the measured value. The highest weight is given to the least uncertainty value. The estimates generated by the method tend to be closer to the actual values than the original measurements, since the weighted average presents a better estimate of uncertainty than both values used in its calculation. From a theoretical point of view, the Kalman filter is an algorithm for efficiently making accurate inferences about a linear dynamic system, which is a Bayesian model similar to a Markov hidden model, but where the state space of the variables is not observed is continuous and all observed and unobserved variables have normal distribution (or often multivariate normal distribution).

c. Fourier Transform

This technique applies mainly to stationary periodic signals.

The Fourier transform is an integral transform that expresses a function in terms of sinusoidal base functions, i.e. as the sum or integral of sinusoidal functions multiplied by coefficients. There are several directly related variations of this transform, depending on the type of function to transform. The Fourier transform decomposes a temporal function (a signal) into frequencies, just as a musical instrument string can be expressed as the amplitude (or volume) of its constituent notes. The Fourier transform of a temporal function is a complex frequency value function whose absolute value represents the sum of the frequencies present in the original function and whose complex argument is the phase of displacement of the sinusoidal base at that frequency.

d. Wavelet Transform

This technique applies mainly to non-stationary Periodic signals.

Many of the time series exhibit non-stationary behaviors such as changing trends, structural breakdowns from the beginning to the end of the event. These features are often the most important parts of the signal and by applying TF it is not possible to efficiently capture these events.

The wavelet transform is a very useful tool for analyzing these non-stationary series.

The wavelet transform has attractive qualities that make it a very useful method for time series, exhibiting characteristics that could vary both in time and frequency (or scale).

The wavelet transform allows the signal to be decomposed into a set of function bases at different resolution levels and localization times. From these levels it is possible to reconstruct or represent a function, using the wavelet bases and coefficients of these levels appropriately.

e. ARMA (p,q) model

This technique applies mainly to stochastic signals.

In the statistical analysis of time series, autoregressive moving average (ARMA) models provide a poor description of a weakly stationary stochastic process in terms of two polynomials, one for autoregression and one for average mobile.

Given a time series of X data, the ARMA model is a tool for understanding and perhaps predicting future values in this series. The model consists of two parts, an autoregressive part (AM) and a moving average part (AM). The AR part involves returning the variable to its own lagged, that is, past values. The AM part involves modeling the error term as a linear combination of error terms that occur contemporaneously and at various times in the past.

The model is generally referred to as the ARMA (p,q) model, where p is the order of the autoregressive part and *q* is the order of the moving average part.

In signal processing, a time series is a collection of observations made sequentially over time. In linear regression models with cross-section data the order of observations is irrelevant to the analysis, in time series the order of data is fundamental. A very important feature of this type of data is that neighboring observations are dependent and the interest is to analyze and model this dependency.

Within probability theory, a stochastic process is a family of random variables representing the evolution of a value system over time. It is the probabilistic counterpart of a deterministic process. Instead of a process that has a single way of evolving, as in the solutions of ordinary differential equations, for example, in a stochastic process there is an indetermination: even if one knows the initial condition, there are sometimes infinite directions in which the process can evolve.

In discrete time, as opposed to continuous time cases, the stochastic process is a sequence of random variables, such as a Markov chain. The variables corresponding to the various times may be completely different, the only requirement being that these different values are all in the same space, that is, in the contradiction of the function. One possible approach is to model random variables as random functions of one or more deterministic arguments, in most cases, relative to the time parameter. Although the random values of a stochastic process at different times seem to be independent random variables, in the most common situations they exhibit a complex statistical dependence.

#### *2.3.2.2 Diagnosis*

At this stage the type and degree of variation, the type of failure, its location, its severity and the incidence on the performance of an element or component are determined, and most particularly the root cause of this disturbance is identified [8].

In order to achieve the objectives of the fault diagnosis stage, two very important techniques allow the monitoring and supervision system to achieve this intelligent and autonomous decision-making system feature [5, 7, 14]. The application of these two methodologies is the major differential between classical monitoring and supervision and intelligent monitoring and supervision. These two methodologies are described below:

1. Fault detection methods

They detect faults, locate them and determine their degree of severity. The methodologies used in this technique to detect faults are:

a. Fault detection with limit checking

It is a simple and often used method to detect faults by checking the limit of a directly measured Y(t) variable. The measured variables of a process are monitored and checked if their absolute values or trends exceed a threshold. An additional possibility is to check its plausibility.

To detect failures in a device of a production system it is necessary to establish or determine variation limits for the variables considered of interest. Usually these limits are the maximum values these variables can reach. When the values of the variables reach these limits, it can be inferred that a variable is presenting abnormal changes. If these changes are continuous or discrete, it can be concluded that the device is in a process of failure.

The fault detection threshold verification technique is based on two procedures in order to achieve its goal, namely:

- Binary thresholds;
- Diffuse thresholds.

In most situations the binary decision between "normal" and "disturbance" is sometimes artificial, because there is rarely a marked difference between these two states. Therefore, the diffuse threshold procedure is a more realistic alternative for detecting changes in the behavior of variables.

2. Fault diagnosis methods—root cause identification.

Identify the impact a failure has on the performance of an element or device. This is strongly related to the severity of the failure. The important part of the diagnostic step is that the system is able to identify the root cause of a problem.

Many measured signals show oscillations that are either harmonic or stochastic in nature or both [9]. If changes to these signals are related to actuator, process and sensor failures, signal model-based failure detection methods may be applied.

Assuming special mathematical models for signal measurement, appropriate characteristics can be determined. Comparison with observed characteristics for normal behavior provides changes in these characteristics which are considered as analytical symptoms.

The signal model can be divided into non-parametric models, such as frequency spectrum or correlation functions, and into parametric models as amplitudes to distinguish frequencies or ARMA models.

At this stage usually the signals that are analyzed for being frequent in different production processes are focused on the following types of signals:

- Periodic signs;
- Non-stationary periodic signals;
- Stochastic signals.

## 2.4 Applying corrective actions: alternatives

Within the various alternatives available to eliminate these changes in the behavior of a given parameter, the best is sought from the technical and economic point of view and, most importantly, that can be applied in real time, either in on mode online or offline [15]. This depends on the severity of the failure, for example whether it is a failure that is likely to happen and the serious consequences or symptoms of an incipient failure.

This optimized solution is found by applying intelligent optimization techniques such as:

- Expert systems;

- Neural networks;
- Fuzzy logic.

An ES is capable of processing non-numerical information, presenting conclusions on a certain subject as long as it is properly oriented and "fed". Another common feature in expert systems is the existence of an uncertain reasoning mechanism that allows one to present uncertainty about domain knowledge. In other words, ES employ human knowledge to solve problems that require the experience of one or more specialists. Within the application of expert systems it is necessary to count on the participation of a robust database, where the knowledge base will be stored (expert knowledge to solve numerous problems).

### *2.4.1 Knowledge base*

The knowledge base is a permanent but specific element of an expert system. This is where the information of an expert system is stored, that is, the facts and rules. Information stored in a particular domain makes the system an expert in that domain.

### *2.4.2 Blackboard*

Communication of information between expert systems is done by a mechanism called a blackboard. A blackboard is a place within computer memory where information stored in an expert system is "pinned" so that any other expert system can use them if you need the information contained therein to achieve your goals.

The blackboard is a structure that contains information that can be examined by cooperative expert systems. What these systems do with this information depends on the application.

Still, a blackboard, draft, or working memory (temporary memory) has a useful life during the course of a query and is linked to a concrete query. It is an area of memory used to make evaluations of the rules that are retrieved from the knowledge base to arrive at a solution.

Information is recorded and erased in an inference process until the desired solution is reached.

### *2.4.3 Inference machine*

Inference engine or inference engine is a permanent element that can even be reused by various expert systems. It is the party responsible for seeking knowledge base rules to be evaluated, directing the inference process [13]. Knowledge must be prepared for good interpretation and objects must be in a certain order, represented by a context tree.

Basically the inference engine is divided into the following steps:

- Select and search;
- Evaluate and Search.

Summarizing the above tasks, it can be said that the rules necessary to reach a goal must be sought in the knowledge base. These rules will be placed on the blackboard, and existing rules will only be evaluated after the most recent ones.

The evaluation order on the blackboard follows a stack-like structure to achieve the most recent goal. The rule will continue to be evaluated as long as the assumption conditions are true, otherwise the rule will be dropped, the set goal will be unstacked and a new rule will be loaded.

When a value of a parameter in a given context is not known and is not in the stack structures, one should then look for new information in the knowledge base, search for new rules, or ask the user directly.

### 2.4.4 Neural networks

An artificial neural network is made up of several processing units whose operation is quite simple. These units are usually connected by communication channels that are associated with a certain weight. Units perform operations only on their local data, which is input received by their connections. The intelligent behavior of an Artificial Neural Network comes from interactions between network processing units.

Neural networks allow optimized selection of a particular solution alternative for a given event or change. The neural network is used in this process to evaluate the results of the expert system, that is, the final solution should be selected as the best of all presented, the neural network allows to establish this solution.

### 2.4.5 Fuzzy logic

Fuzzy logic is based on fuzzy set theory. Traditionally, a logical proposition has two extremes: either it is completely true or it is completely false. However, in Fuzzy logic, a premise varies in degree of truth from 0 to 1, which leads to being partially true or partially false.

Fuzzy logic is the logic that supports the modes of reasoning that are approximate rather than exact. Fuzzy systems modeling and control are techniques for rigorously handling qualitative information. Derived from the concept of fuzzy sets, fuzzy logic forms the basis for the development of process modeling and control methods and algorithms, reducing the complexity of design and implementation, making it the solution to control problems hitherto intractable classic techniques.

In classical and modern control theories, the first step in implementing process control is to derive the mathematical model that describes the process. The procedure requires knowing in detail the process to be controlled, which is not always feasible if the process is too complex. Existing control theories apply to a wide variety of systems where the process is well defined.

However, all of these techniques are not capable of solving real problems whose mathematical modeling is impractical. For example, in many situations a considerable amount of essential information is only known a priori qualitatively. Similarly, performance criteria are only available in linguistic terms. This picture leads to inaccuracies and inaccuracies that make it impossible to use most of the theories used so far.

Fuzzy modeling and control theory are techniques for rigorously handling qualitative information. It assesses how imprecision and uncertainty should be managed and, in so doing, become powerful enough to properly manipulate knowledge. This technology considers the relationship between inputs and outputs, aggregating various process and control parameters. This allows processes considered complex to be reconsidered so that the resulting control systems provide a more accurate result as well as stable and robust performance. The sheer simplicity of implementing fuzzy control systems can reduce the complexity of a project to a point where previously intractable problems are now solvable.

## 2.5 Solution

The intelligent monitoring and supervision system becomes autonomous to decide, but this decision should indicate the best action that should be taken to mitigate or eliminate a particular change. This intervention must be in real time and in online mode. It should show the type of application that will be performed, the point where the intervention will be made, the components that will be reached and their intervention time. This solution can be given by the following procedure:

### *2.5.1 Maintenance*

Depending on the situation, it can be in real time and in Online mode, meaning the maintenance team can make the necessary adjustments without shutting down the equipment and reducing its availability in the shortest possible time. The smart system should provide recommendations for making these correctives without compromising equipment operation. This is accomplished through expert system intervention. This system will decide on what type of maintenance to perform.

### *2.5.2 Predictive and proactive maintenance*

Depending on the situation, it can be in real time and in Online mode, meaning the maintenance team can make the necessary adjustments without shutting down the equipment and reducing its availability in the shortest possible time.

The smart system should provide recommendations for corrective action without compromising equipment operation. This is accomplished through expert system intervention.

### *2.5.3 Element or device replacement*

The intelligent system must have the ability to make this decision, supported by technical and economic criteria (losses).

## 2.6 Innovation

The final and ultimate solution will demonstrate the versatility, autonomy and efficiency of monitoring and supervision systems when they work and are structured as intelligent systems. Errors in procedures with these systems will be minimal to ensure safety and accuracy.

## 2.7 Benefits

Safe (high availability and reliability) and efficient operation of production systems, balanced and timely investments, reduced operating and energy costs.

Following is a typical output (report) of the software developed for the implementation of the methodology, especially the part related to data evaluation, fault diagnosis, root cause finding and determination, action decision making to be performed and execution of these actions [3, 4].

Following is an output of the software developed to implement the proposed intelligent system framework [1, 15]. All the tools presented in this chapter have been included:

**Sensor: 45**
Transmitted signal: Variation of field current

**EVALUATION**
Failure Mode: Sudden and Continuous Field Current Change
Failure Mode: Voltage Regulator Saturation
Behavior: Machine Rotor Vibration

**DIAGNOSIS**
Sudden and continuous change of the magnetic field of the generator.
Loss of machine efficiency
Machine capacity reduction
Reduction of magnetic field in some windings.
Fault Type: Mechanical and Electrical

**ROOT CAUSE**
Increased resistance of the machine field windings.
Loosening of the connection between magnetic coil field coils.

**ACTIONS TO BE CARRIED OUT**
Make the machine unavailable for operation.
Change the connection system between magnetic coil field coils.

## 3. Conclusions

Intelligent monitoring and control systems allow minimize the risks of failure of production systems, as is the case with generation systems (will be taken as a reference for this project), where intelligent systems are widely used, especially in large and top technology plants, and consequently increase its reliability (reduced failure rate/year) and the availability, improving the quality of energy supply by reducing the periods and interruption frequency of power supply, by improving indicators DEC, FEC, DIC, FIC and DMIC, and reliable management charge and distributed generation.

Centralization of information processing by intelligent monitoring systems will improve the efficiency of operation of electrical systems, optimize maintenance processes within the generation plants and consequently increase or maintain the estimated useful life of the generators, economically benefiting utilities power.

The transformation of the current systems for monitoring and supervision of hydroelectric plants in intelligent systems, effectively represents a technological advancement over conventional monitoring systems. What defines the quality of the response of these systems, in relation to the supervision and diagnosis, it is the experience of those responsible for analysis of failure modes.

The data management infrastructure, established by the power utilities, will be responsible for more or less extracted benefit of the system as well as for maintaining the efficient operation of the same. The choice of the best strategy for the Data and Information Management, will depend on the policies adopted by companies to their treatment. The benefits of smart grid technology, in monitoring systems will come when there is a data management policy functional within the utilities should be avoided as much as possible "excess of monitored parameters". Prioritization criteria of failures "detectable" or "observable" must be considered. For each observable failure mode, there will always be a form of detection, which keeps a relationship "sensitivity/installation cost" more, and that in principle, should be chosen.

An action of great interest, which should be considered in the monitoring and supervision systems is the integration of auxiliary systems, to conduct their analysis and diagnosis, together with those from the main systems, causing minimal impact on its cost of installation. The influence of failures in auxiliary systems (ancillary services) with the probability of generating, forced stops of the equipment and system is high and in some situations similar to those of the main systems.

Importantly, there is not a single application and solution of systems or smart grids. Many of these functions will not become viable if coexist with others and should be implemented according to the needs of utilities. Thus, individual functions such as monitoring and fault detection in generators or feeder circuits, may not have their benefits evaluated separately.

## References


[1] Selak L, Butala P, Sluga A. Condition monitoring and fault diagnostics for hydropower plants. Computers in Industry. 2014;**65**(6):924-936. ISSN 0166-3615

[2] Perez GA, Nelson K. Integration of distributed generation in power distribution networks and its structure as an intelligent generation system. In: 2015 IEEE PES Innovative Smart Grid Technologies Latin America (ISGT LATAM). Montevideo, Uruguai. pp. 134-139

[3] Working Group A1.11. Guide for On-Line Monitoring of Turbogenerators. CIGRE; 2010

[4] Working Group A1.10. Survey of Hidrogenerator Failures. CIGRE; 2009

[5] Wenye W, Yi X, Mohit K. A survey on the communication architectures in smart grid. Computer Networks. 2011;**55**(15):3604-3620

[6] Wei D, Lu Y, Jafari M, Skare P, Rhode K. An integrated security system of protecting Smart Grid against cyber attacks. Innovative Smart Grid Technologies (ISGT). United States, Gaithersburg, MD; 2010:1-7. Available from: https://doi.org/10.1109/ISGT.2010.5434767. INSPEC Accession Number: 11205470

[7] Xiang L, Wenye W, Jianfeng M. An empirical study of communication infrastructures towards the smart grid: Design, Implementation and evaluation. IEEE Transactions on Smart Grid. 2013;**4**(1):170-183

[8] Isermann R. Fault Diagnosis Systems—An Introduction from Fault Detection to Fault Tolerance. Berlin, Heidelberg, New York: Springer; 2009. ISBN-10 3-540-24112-4. ISBN-13 978-3-540-24112-6. Library of Congress Control Number: 2005932861

[9] Ding SX. Model-Based Fault Diagnosis Techniques. Berlin, Heidelberg: Springer-Verlag; 2009. eBook ISBN: 978-3-540-76304-8; DOI: 10.1007/978-3-540-76304-8

[10] Helio PA Jr, Levy AFS, Carvalho AT. Estudo sobre a Influência dos Acopladores Capacitivos na Sensibilidade da Medição de Descargas Parciais em Máquinas Elétricas Rotativas. In: XX SNPTEE, Recife, Brasil. 2009

[11] Omori J. O projeto de Smart Grid da COPEL. In: Smart Grid Brazil Forum, São Paulo. 2010

[12] Piirainen J. Applications of horizontal communication in industrial power stations [Master in science thesis]. Tampere University; 2010

[13] Xin Y, Baldine I, Chase J, Beyene T, Parkhurst B, Chakrabortty A. Virtual smart grid architecture and control framework. In: IEEE Conference Publications (SmartGridComm). IEEE; 2011. pp. 1-6

[14] Wenye W, Zhuo L. Cyber security in the smart Grid: Survey and challenges. Computer Networks. 2013;**57**(5):1344-1371

[15] Guimaraes PHV, Murilo A, Andreoni M, Mattos DMF, Ferraz LHG, Pinto FAV, et al. Comunicação em redes elétricas inteligentes: Eficiência, Confiabilidade, Segurança e Escalabilidade. In: Minicursos do Simpósio brasileiro de redes de computadores—SBRC-2013, Brasília, DF, Brazil. 2013. pp. 101-164

Chapter 8

# An Approach to Optimize the Management of Information Security in Public Organizations of Ecuador


## Abstract

The problems of information security in public organizations in Ecuador are evident, which, as a result, have led to corruptions that are present at all levels of operational, tactical and strategic management. The objective of this chapter is to analyze the available information found in different media, written, spoken, among others. The deductive method was used for the collection of information and observation techniques. It turned out the improve in the administrative processes, prototype diagram of sequence of access of users and services, prototype of integration of technologies of security of the information for public organizations of Ecuador. It was concluded that to avoid corruption in a country change should happen at all levels: the way of thinking and culture of the inhabitants, laws, penalties to politicians without parliamentary immunity, application of information and communications technologies (ICT) in an appropriate manner, and complying with international standards in information security. To improve information security, administrative policies on information security must be changed, and technologies related to immutable security algorithms, Ledger, Hyperledger, etc., must be used.



**Keywords:** information security, information security management, database security, public organizations of Ecuador, security models, cryptography


## 1. Introduction

Public organizations in Ecuador have problems in the management of Information Security. The "Ministry of Telecommunications and the Information Society" ratify that information security problems persist. According to the publication of the "White Book of the Information and Knowledge Society", it turned out that only 8% comply with the Security Policies, and those responsible for information security that are part of IT have 51% and that are part of Contingency Plan only have 16%, among other security indicators [1].

The company Deloitte conducted a study in 2017 concerning the problem of information security, and the results were published by the "White Paper on the Information and Knowledge Society," in which more than 50 national and multinational companies participated to improve information security management and the following was determined:

1. Around 50% had some security breach, and of this, 20% could not determine the impact of this gap, since they did not have an incident management process.

2. Nearly 50% indicated that their main initiative for 2018 will be training and awareness in information security.

3. More than 50% cited as one of their main difficulties the lack of budget, followed, very closely, by aspects such as the lack of visibility and influence and the lack of competent personnel.

4. Around 75% did not measure the return on investments in information security.

5. The 20% were prepared to face security incidents, originated in social networks.

6. The 60% did not have an SOC (Security Operation Center); meanwhile, almost 20% said they will have one by 2018.

7. The 36% did not have a disaster recovery plan.

8. As a result of internal and external reviews of companies, user management remains the most shaky element in the management of CISOs (Chief Information Security Officer (Deloitte, 2017) [1].

Among others defined by the CENDIA published in 2017 that is recorded in the "White Paper of the Information and Knowledge Society".

The implementation projects of the Information Security Management System ISO 27000 ensure all the information assets to have complete control of the organization according to what is stated in the book "Public Companies and Planning" [2].

The security of information is critical today in all public or private organizations; based on this reason, it is necessary that Latin and world universities generate specialized careers in the area of information security to provide qualified personnel considering that information security is a key aspect for the management of an organization [3].

With the foregoing, it is confirmed that public and private organizations in Ecuador and in a large part of the world have serious problems of information security. Information is considered as data, videos, sound, and documents, among others, that can be saved, shared, socialized, etc.

Therefore, mishandling of information can lead to failure of organizations; on the other hand, correct decisions can be made based on information that provides confidentiality, integrity, and authenticity.

In accordance with the current paradigms in information security and computer auditing, the following most relevant points to be considered by public organizations to improve information security management were determined:

1. Change of information security culture in first level executives, so that they consider that information security is not a cost, but is an investment to guarantee the mission, vision, and strategic objectives of an organization.

2. All persons working in public organizations, both the first authority and the lowest office, which may be the custodian or guard, are an important

and responsible party in order to maintain the integrity of the information. One of the main errors in the management of information security in public companies is that we are convinced that only those who handle information or strategic managers are responsible.

3. The structural and functional organizations currently available to public organizations do not allow information and communications technologies (ICT) coordinators/directors/managers to govern the organization.

4. The lack of planning and control in a globalized way for the generation of security plans, contingency, backup, and protection against natural disasters, etc., causes vulnerabilities, risks, and threats in the security of information in the organization.

5. Adequate security models and technologies are required for each public organization considering the mission, vision, and strategic objectives.

6. There should be qualified personnel with experience with an average 10 years in the area of information security and with academic training at all levels (Engineering, Master's, and Doctorate) in the same area of knowledge in accordance with the provisions of UNESCO, SENESCYT, CES (title nomenclature) [4].

7. The World Bank determines that one of the main causes for corruption in Latin America and the Caribbean is that there is no adequate management of Information and Communications Technologies (ICT) in the area of Information Security, and proposes to use it as technologies' alternative such as blockchain, Ledger, and Hyperledger. It also clarifies that as long as there is direct human intervention in the processes and no adequate technologies are used, there will be a greater probability of corruption and the only ones who pay for this incorrect management will be the low-income inhabitants [5].

## 2. Security of the information

Information security worldwide is considered the main fixed asset of a public or private organization. With the appropriate management of information, corruption in public-private organizations can be avoided, such as transfers of money without due authorization, terrorist attacks, information theft, manipulation of processes and legal reports, kidnappings, violations, accidents, prevention of natural disasters, etc.

### 2.1 Pillars of information security

To carry out the analysis of information security, the current situation and the functions of the security pillars must be considered clearly: vulnerabilities, risks, threats, which will have a direct relationship with the identity, authenticity, authorization, and audit (IAAA), so that the information is with confidentiality, integrity, and availability (CIA).

The following is a structure for the security of information in public organizations of Ecuador, considering the pillars of security to mitigate the vulnerabilities, threats, and risks of information.

**Figure 1** shows that information systems have vulnerability, threats, and risk generation and have two layers of security that are covered first with the identity,

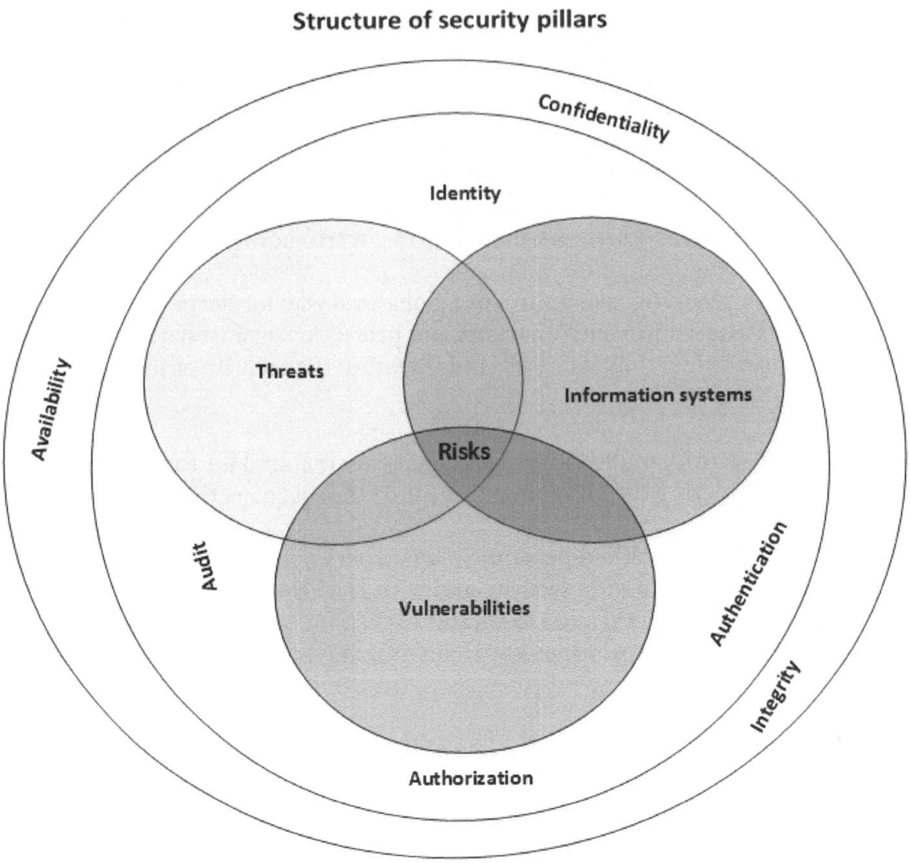


**Figure 1.**
*Pillars of information security.*

authentication, authorization, and audit (IAAA) and the second with the confidentiality, integrity, and availability (CIA).

### *2.1.1 Analysis of information security pillars*

Public organizations in Ecuador have vulnerabilities, and as a result, threats and risks are generated for not having adequate procedures for users to have identity, authenticity, authorization, and audit (IAAA), so that the delivery of the information to internal and external users is with confidentiality, integrity, and availability (CIA), then the general description of the following pillars of security will be made.

- Identity is considered to internal or external users who have access to information.
- Authentication corresponds to the identification of users for access through technological or manual system.
- Authorization corresponds to what information the user who has been identified and has an authentication is entitled.
- Audits are the processes and activities performed in the user and recorded in a log to store their identity, authentication, and authorization, to be used at any time in processes of computer audits.

- Confidentiality of the information is considered to be the right that guarantees access only to the personnel that previously have authorization under its responsibility.

- Integrity of the information is when the information is not modified from the beginning of its generation until the final delivery to the authorized persons.

- Availability of information, so that it can be used by users; depending on a technological infrastructure, the availability of information can be guaranteed.

*2.1.2 Consequence due to the incorrect management of information security*

To solve the problems of the security information of vulnerabilities, risks, threats, which has a direct relationship with the identity, authenticity, authorization, audit (IAAA) and with the security triangle confidentiality, integrity, availability (CIA), one of the alternatives is to carry out the risk analysis; apply the Cobit 5.0 methodology references adapted to the public organization regarding information security ISO 27001; evaluate the degree of knowledge and implementation of information security management systems, based on the norm NCh-ISO 27001, ITIL, COSO; generate or adopt models and appropriate security technologies for each organization; apply immutable security algorithms; generate or adopt own methodologies of the organization for the change of computer culture; and make plans of security, among others.

*2.1.3 Alternatives to solve information security problems*

To solve the problems of vulnerabilities, risks, threats; which has a direct relationship with the identity, authenticity, authorization, audit (IAAA) and with the security triangle confidentiality, integrity, availability (CIA); one of the alternatives is: Carry out the risk analysis, Apply the Cobit 5.0 methodology references adapted to the public organization regarding information security ISO 27001, evaluation of the degree of knowledge and implementation of information security management systems, based on the norm NCh-ISO 27001, ITIL, COSO, generate or adopt models, appropriate security technologies for each organization, apply immutable security algorithms, generate or adopt own methodologies of the organization for the change of computer culture, make plans of security among others.

Also take as a reference other similar projects such as the one applied in a health institution in Chile [6].

## 3. Related investigations

### 3.1 Publications related to the research topic

Below is a list of the articles published in different conferences and scientific journals directly related to the public organizations of Ecuador, in the area of information and communications technologies (ICT) and information security.

Indicator Model for measuring the Alignment between Institutional Strategies and ICT Strategies for a Public Sector Company [7], Las TIC en el Ecuador [8], Tecnologías de Información y Comunicación Impactan la Optimización de los Procesos para el Desarrollo Local [9], Analysis to define management of identities access control of security processes for the registration civil from Ecuador [10], security analysis of civil registry database of Ecuador [11], an approach to information security by applying a conceptual model of identities in smart cities projects [12], adequate security protocols adopt in a conceptual model in identity management for the civil registry of Ecuador [13], analysis of model Clark Wilson to adopt to the database of the civil registry of Ecuador [14], mitigating the security of the database by applying a conceptual model of integrity for the civil registry of Ecuador [15], a security algorithms approach to apply to the civil registry database of the Ecuador [16], conceptual model for identity management to mitigate the database security of the registry civil of Ecuador [17], adoption of the Hash algorithm in a conceptual model for the civil registry of Ecuador [18], an approach of efficient security algorithms for distribute architectures [19], biometric systems approach applied to a conceptual model to mitigate the integrity of the information [20], algorithms for efficient biometric systems to mitigate the integrity of a distributed database [21], analysis of efficient processes for optimization in a distributed database [22], analysis of HIPAA for adopt in the information security in the civil registry of the Ecuador [23], a blockchain approach to mitigate information security in a public organization for Ecuador [24], analysis of the appropriate security models to apply in a distributed architecture [25], optimization of an electronic signature scheme in a voting system in a distributed architecture [26], ensuring the blind signature for the electoral system in a distributed environment [27], analysis cryptographic for electronic votes in systems of distributed architectures [28], an approach to the efficient security algorithms used in voting scanning in an electoral process [29], a homomorphic encryption approach in a voting system in a distributed architecture [30], analysis of security algorithms for a distributed database [31], a Hyperledger scheme for the deployment of smart contracts in a public organization of Ecuador [32], analysis of adequate bandwidths to guarantee an electoral process in Ecuador [33], appropriate security protocols to mitigate the risks in electronic money management [34], cryptographic algorithms to mitigate the risks of database in the management of a smart city [35], impact on the information security management due to the use of social networks in a public organization in Ecuador [36], an information security approach in the armed forces of Ecuador [37].

### 3.2 General summary of articles published by segments

- The management of information and communications technologies (ICT), Models of Indicators, allows to visualize all the processes and activities in general form of public institutions that must be analyzed with priority to be considered strategic [7–9].

- The analysis of information security regarding models, technologies, conceptual models, security protocols, prototypes, and cryptographic algorithms, among others, for the civil registry of Ecuador 10–25].

- They support the analysis, design, models, and prototypes of security for the National Electoral Council to mitigate the risks in the integrity of the information that will be delivered from the electoral processes [26–33].

- Analysis of appropriate security protocols to guarantee the cash flow of public organizations using electronic money [34].

- Cryptographic analysis that allows improving the security of data in smart cities that involve the main cities of Ecuador such as Quito, Guayaquil, and Cuenca, among others [35].

- Impact of social networks on information security in public organizations, as it affects both internal and external users to prevent the information from being disclosed without control [36].

| Odr. | Public organization name | Priority | | |
|---|---|---|---|---|
| | | Low | Half | High |
| 1 | National directorate of the civil registry of Ecuador | | | x |
| 2 | National electoral council | | | x |
| 3 | National directorate of public data | | | x |
| 4 | National council of the judiciary | | | x |
| 5 | National assembly | | | x |
| 6 | General comptroller of the state | | | x |
| 7 | Joint command of the armed forces | | | x |
| 8 | Secretaria de gestión de riesgos | | | x |
| 9 | National secretary of higher education science, technology and innovation (SENESCYT) | | | x |
| 10 | Central bank of Ecuador | | | x |
| 11 | National secretary of planning and development, Ecuador | | | x |
| 12 | Ministry of communications and the information society | | | x |
| 13 | Superintendency of banks | | | x |
| 14 | Internal revenue services of Ecuador among others | | | x |

**Table 1.**
*Main organizations of Ecuador that should be evaluated in the first phase.*

- The security of information in the joint command of the armed forces of Ecuador is important to analyze because it guarantees the internal and external sovereignty of the country [37].

### 3.3 Priority of Ecuador's public organizations for the analysis

It defines public institutions that have problems in the management of information and communications technologies (ICT) and information security [1]. For this analysis, public organizations with the highest priority for the evaluation of information security management are considered; the same that in the medium term should be analyzed [7–8]. We must mention that all public organizations in Ecuador must improve the management of information security, but it should be done in phases.

In **Table 1**, the public organizations that should be in the first phase are detailed, considering that they have a high priority for the interrelation they have in state processes, to reduce corruption with the use of appropriate information and communication technologies.

## 4. Models, security technologies, and good practices

To apply the models, security technologies, and good practices, the mission, vision, and strategic objectives of each organization must be analyzed; it is not appropriate to apply the same to everyone, given that each organization will have its priorities for the application of confidentiality, integrity, and availability of information. Each model, technologies, and good practices have different strengths, which can be applied appropriately [38].

## 4.1 Security models

### *4.1.1 Model Clark-Wilson*

The Clark-Wilson model is based on four principles: authentication, audit trail, separation of obligations, and well-formed transactions.

Clark-Wilson is a widely used model to protect business information against unauthorized modification. In the CW model, the data in the system are requested in two groups:

- Restricted data elements (CDIs) that are elements or objects whose integrity must be maintained
- Unrestricted data elements (UDIs) that are elements or objects that are not covered by the integrity policy, such as the input data, but which are relevant since they can be transformed into CDIs [39]

### *4.1.2 Chinese Wall Model*

The Chinese Wall model is oriented to guarantee the confidentiality of the information it raises and provides controls to reduce conflicts of interest that may exist between organizations that handle the same business logic [40].

### *4.1.3 Model Bell-LaPadula*

The Bell-LaPadula model's strength lies in multilevel security, which does not allow sensitive information to be filtered by people or entities that do not have the appropriate level of access; this helps maintain a certain degree of confidentiality [41].

## 4.2 Security technologies

### *4.2.1 Cryptography*

The importance of cryptography is that it is the only current method able to enforce the objective of computer security "maintain privacy, integrity, and authenticity" and enforce nonrejection, related to not being able to deny authorship and reception of a message sent [42].

### *4.2.2 Log inmutables*

Applications require robust and inviolable registration systems, for example electronic voting or bank information systems. At Scytl, we use technologies called immutable records, which are implemented in electronic voting solutions. This technology ensures the integrity, authenticity, and nonrepudiation of the generated records; therefore, in case of any event, the auditors can use them to investigate the problem. To improve the integrity of the information, an implementation for immutability is required, the integrity tests of the secure registers within the chain of blocks known as Bitcoins that is based on SHA-1 [43].

### *4.2.3 Biometric systems*

Biometrics is considered a solution in information security problems; biometrics has the necessary assurances that the information stored in databases in institutions cannot be manipulated and lose their integrity. There are some types of biometric systems that can be used, such as fingerprint, iris reader, facial recognition, and voice recognition. The use of multimodal biometrics has been considered in the study [44].

*4.2.4 Ledger*

The Ledger technology is based on the blockchain that ensures the registration of information in a distributed architecture, at the highest possible level, despite being distributed. With Ledger technology, we are thinking of a specific purpose distributed network, a network that shares a local maintenance [45].

*4.2.5 Hyperledger*

It is a technology that consists of a network infrastructure based on blockchain. Hyperledger fabric (HLF) is an open source implementation of a distributed accounting platform to execute intelligent contracts in a modular architecture. The implementation of Hyperledger technologies will mitigate the risks of information; because in all the transactions you make, you will register through immutable log [46].

### 4.3 Good practices

All public organizations must apply good information security practices such as those defined in ISO 27001, define appropriate indicators, change of culture in the area of information and communications technologies (ICT) by executives, consider the Cobit 5.0 methodology that the technologies of information and communications govern the organization (separate what is management and government).

In addition, the following good practices are suggested: update systems, limit users, block output systems, separate the most important files, automate, monitor permanently, define safety standards, unify processes, and educate internal and external users.

## 5. Alternatives to improve information security

In this research, several alternatives were analyzed to improve the security of the information such as mechanical safety that is applied in an appropriate way for each organization and definition of all the processes of each public organization, models, prototypes, and cryptographic security algorithms using techniques of flow chart, etc. [7–33].

### 5.1 Description in general

It should be noted that in the publications of the reference of [7–33], you can find the following information:

The situation of Information and Communications Technologies (ICT) in Ecuador, definition of processes, conceptual security models, cryptographic algorithms, security models, analysis of security protocols, technological infrastructures, technologies, applied to public organizations in Ecuador in this case to the Civil Registry and National Electoral Council of Ecuador. To be considered as alternatives to improve the management of information security.

## 6. Proposals to improve information security

After having analyzed all the published articles [5–31] that are directly related to this chapter of the book, the following activities to improve the management of information security are proposed to be carried out.

### 6.1 Improve administrative processes

In administrative processes, it is important that planning bodies modify the functional structural organization of public organizations, considering the good practices of Cobit 5.0. In the organic structure, the general manager (CEO) and the information and communications technologies (ICT) manager (CIO) must be at the same level.

**Figure 2** defines a generic structure chart suitable for public organizations where the manager/coordinator/director of information and communications technologies (ICT) can govern the organization to comply with the recommendations of Cobit 5.0.

#### *6.1.1 Change of culture in information and communications technologies (ICT)*

To make the change in the information and communications technologies (ICT) culture, a training plan is required, with an appropriate methodology at all operational, tactical, and strategic levels, especially at the strategic level so that they are clear. For an organization to be competitive and the management of information security to improve, information and communications technologies (ICT) must govern public organizations.

This change of culture at the level of high-level officials of public organizations is necessary to execute, considering that 95% of the authorities of public organizations defined by information and communications technologies (ICT) at the operational level are convinced that they are simply a support for the management of the organization [7].

#### *6.1.2 Processes and activities that should be considered*

To carry out this activity, it can be executed through different types of indicators in the information and communications technologies (ICT) area; in this case, the following indicators are used as an alternative: Degree of Utilization, Degree of Support for the Process, Degree of Use, Degree of Online Support,

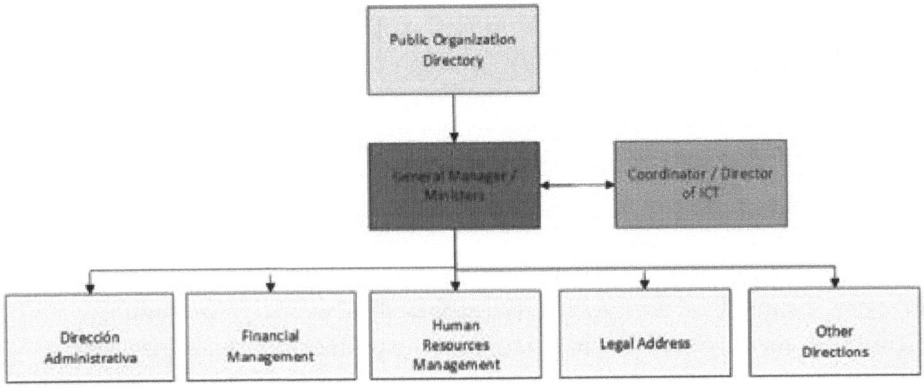


**Figure 2.**
*Generic structural organization chart of a public organization.*

**Figure 3.**
*Frequently asked questions on information security.*

Degree of Scope, Degree of Coverage, Degree of Operational Support, Degree of Management Support, and Degree of Support Corporate. This allows to determine the current situation of the organization in all areas. With this information, it is more feasible to identify the information security situation to improve its management [7].

Another alternative to get to identify the current situation of information security is to ask the following frequently asked questions:

In **Figure 3**, the frequent questions are asked with the objective that during this process the entry of the information is determined and also the output of the information with identity, authenticity, authorization and audit (IAAA) and confidentiality, integrity and availability (CIA).

### *6.1.3 Phases that must be considered to improve the security of information*

Consider the results obtained in the different articles published on the public organizations of Ecuador, to consider as an alternative with the objective of improving the security of information [7–33].

1. Adopt or generate a training plan with appropriate methodology for the public organization for the change of computer culture at the operational, tactical, and strategic levels.

2. Perform the analysis and define the organizational structure of the organization considering the Cobit 5.0 methodology as a reference, where the general manager (CEO) and the manager/coordinator/director of information and communications technologies (ICT) equivalent to chief information officer (CIO) have the same level of authority and the CIO is the one who governs the organization.

3. Carry out the analysis to define the vulnerabilities, risks, and threats that are generated.

4. Define the structure for the execution of the project: general coordinator, specialist in information security, process specialist, administrator of information and communications technologies (ICT) infrastructures, etc., all with academic training in the area of knowledge at all levels, engineering, masters, and if the case deserves in the doctorate fulfilling standards of the SENESCYT and UNESCO. Also have a referential budget.

5. Consider the application of ISO 27001:2013 regarding the certification process by FIRST (International Incident Management Community, CSIRTS, and CERTS).

6. Take into account the good practices of the Cobit 5.0, ITIL, and COSO methodologies to integrate information security management in a globalized manner.

7. Define functions for a work group.

8. Define the steps to be taken with the respective responsibility to each officer of the organization.

9. Prepare an information technology strategic plan (ITSP) with their respective security plans, contingency, backup, etc., with the participation of official's at all operational, tactical, and strategic levels.

10. The exposed phases are those that are suggested to be used as an alternative to define a generic methodology for public organizations in the next chapter of the book.

## 6.2 Prototype sequence diagram of users and services access

A prototype of the sequence of accesses of internal/external users and services to the information of public organizations is defined.

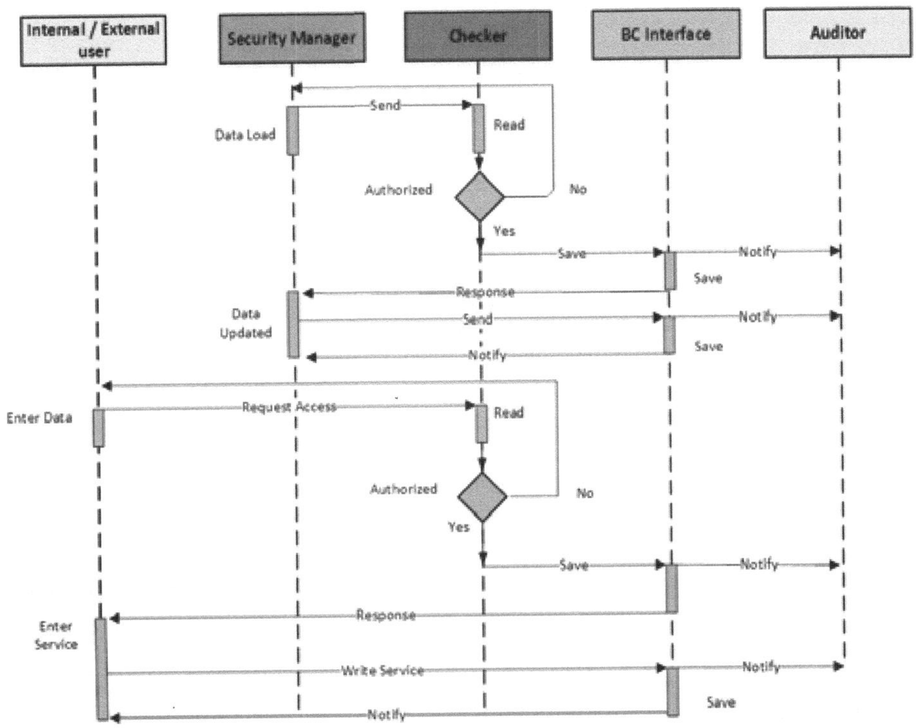


**Figure 4.**
*Prototype sequence diagram of users and services access.*

**Figure 4** describes the dynamic interaction of the user, administrator, verifier, interface, and auditor to perform the sequence of access to information and services.

## 6.3 Prototype of security technology integration

In order to generate this prototype, the recommendations of the World Bank for Latin America and the Caribbean are taken into account, which states textifically: "In the midst of all the technological advances we are currently experiencing, Blockchain, or chain of blocks, has the power to alter current models economic and business, and become a particularly valuable asset for emerging economies. According to the experts, it could also be very useful as a method to fight corruption, especially in Latin America and the Caribbean, where the penetration of mobile telephony can facilitate technological adoptions" [5], the application of Hyperledger Fabric technology with intelligent contracts, provides a starting point to understand a chain of Hyperledger Fabric blocks to consult and update with a Ledger to generate X.509 certificates that are used by applications that interact with a blockchain authorized [47].

**Figure 5** describes how the Hyperledger Fabric, which is a set of functions, uses the Ledger to initiate the status information and the read/write requests through the connections.

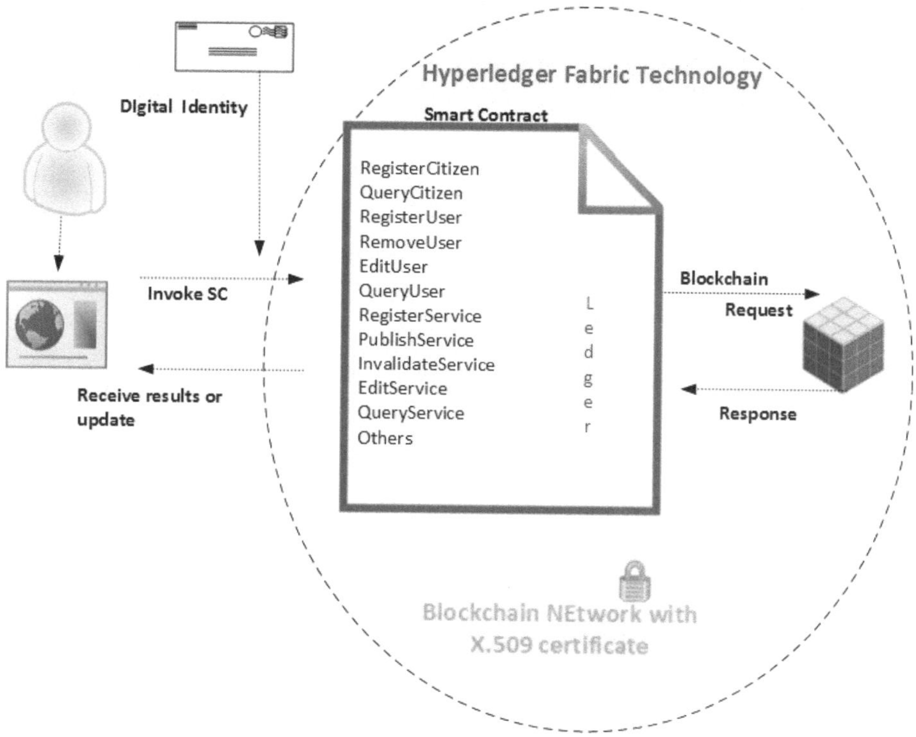


**Figure 5.**
*Prototype integration of security technologies.*

## 7. Conclusions

The security of information is considered strategic and the main asset of public and private organizations. In this chapter, we consider the previous analyzes carried out by the authors in the area of information and communications technologies (ICT) and information security to determine the impact it has on the incorrect management of information. The weakness in the administration of information security is taken advantage of by all officials or workers at an operational, tactical, and strategic level of the public organizations of Ecuador to generate corruption such as incorrect identification of citizens, dead voters, embezzlement in public coffers, false titles especially acquired by the politicians, for all the aforementioned it is concluded that to avoid corruption in a country at all levels the first matter that we are the inhabitants must be changed, laws with strong sanctions to the politicians without parliamentary immunity, application of information and communications technologies (ICT) in suitable form, to fulfill international standards in administration of security of the information like ISO 27001, Cobit 5.0, definition of profiles for the selection of information and communications technologies (ICT) managers, directors or coordinators who are from the area of knowledge in undergraduate degrees, masters, and doctorates, complying with the provisions of SENESCYT and UNESCO. That planning organisms such as the National Secretariat of Planning and Development-Ecuador (Senplades) and the Ministry of Labor Relations consider the position of manager, director, general coordinator of information and communications technologies (ICT) at the same level as the main authority of the Public Organization, Change of culture in the directive civil servants who must consider that the information and communications technologies (ICT) are those that a public organization must govern to be competitive.

To improve the security of information, administrative policies must be changed in information security, using technologies related to immutable security algorithms, Ledger, Hyperledger, etc..

## References

[1] Santacruz GL. Libro Blanco de la Sociedad de la infomración y del Conocimiento. Ministerio de Telecomunicaciones y Sociedad de la Infomración (MINTEL). 2018. Available from: https://www.telecomunicaciones.gob.ec/wp-content/uploads/2018/07/Libro-Blanco-de-la-Sociedad-del-Información-y-del-Conocimento.pdf [Accessed: April 08 2019]

[2] Campaña JA. Empresas públicas y planificaciòn. Senplades; 2013. Available from: http://www.planificacion.gob.ec/wp-content/uploads/downloads/2014/02/Libro-Empresas-Públicas-web.pdf [Accessed: April 04 2019]

[3] Pacheco F. The need for formal education on information security. IEEE Latin America Transactions. 2013;**11**(1):668-670

[4] Ces. República del ecuador consejo de educación superior rpc-so-24-no.480-2017. no. 289; 2017

[5] Mundial B. Blockchain: cómo asegurarse que cada dólar llegue a quien lo necesita [Online]. 2019. Available from: https://www.bancomundial.org/es/news/feature/2019/01/24/blockchain-como-asegurarse-que-cada-dolar-llegue-a-quien-lo-necesita [Accessed: April 11 2019]

[6] Rienzo A, Ieee M, Bustamante M. Evaluation of the degree of knowledge and implementation of information security management systems, based of the NCh-ISO 27001 standard, in health institutions. In: 2018 IEEE Int. Conf. Autom. Congr. Chil. Assoc. Autom. Control; 2018. pp. 1-6

[7] Sistemas DE. Modelo de indicadores para la medición del alineamiento entre las estrategias institucionales y las estrategias de TIC'S para empresa del sector publico. Escuela Politécnica Naiconal; 2013

[8] Toapanta M. Las TIC en el Ecuador. Revista International Estrategos. 2015;**1**(1):7-19

[9] Maciel R, Toapanta M. Tecnologías de información y comunicación impactan la optimización de los procesos para el desarrollo local. In: I Congreso Internacional Investigación en Desarrollo Local y Emprendimiento Socioeconómico Sustentable y Sostenible; Sección Tecnologías de Información: Guayaquil-Ecuador; Digital 20; 2015. pp. 310-323. ISBN: 978-9942-14-417-1

[10] Segundo S, Toapanta M, Luis PD, Mafla E. Analysis to define management of identities access control of security processes for the registration civil from Ecuador. In: IEEE 2nd International Smart Cities Conference: Improving the Citizens Quality of Life, ISC2 2016-Proceedings; 2016. pp. 1-4

[11] Student PD, Moisés S, Toapanta T, Luis PD, Mafla E. Security analysis of civil registry database of Ecuador. In: Int. Conf. Electr. Electron. Optim. Tech. 2016; 2016. pp. 1024-1029

[12] Moises Toapanta AO, Mafla E. An approach to information security by applying a conceptual model of identities in smart cities projects. Journal of Engineering and Applied Science. 2017;**12**(6):7765-7770

[13] Toapanta M, Mafla E, Orizaga A. Adequate security protocols adopt in a conceptual model in identity management for the civil registry of Ecuador. In: IOP Conf. Ser. Mater. Sci. Eng.; vol. 225; 2017. pp. 1-6

[14] Moisés S, Toapanta T, Enrique L, Gallegos M. Analysis of model Clark

Wilson to adopt to the database of the civil registry of Ecuador. Conf. Open Innov. Assoc. Fruct; vol. 21; 2017. pp. 513-518

[15] Moisés S, Toapanta T, Enrique L, Gallegos M. Mitigating the security of the database by applying a conceptual model of integrity for the civil registry of Ecuador. In: Conf. Open Innov. Assoc. Fruct; vol. 21; 2017. pp. 507-512

[16] Moisés TT, Antonio OT, Enrique MG. A security algorithms approach to apply to the civil registry database of the Ecuador. In: IEEE CITS 2017-2017 International Conference on Computer, Information and Telecommunication Systems; 2017. pp. 287-290

[17] Toapanta M, Mafla E, Orizaga J. Conceptual model for identity management to mitigate the database security of the registry civil of Ecuador. Materials Today: Proceedings. 2018;**5**(1):636-641

[18] Toapanta M, Mafla E, Orizaga A. Adoption of the Hash algorithm in a conceptual model for the civil registry of Ecuador. In: AIP Conference Proceedings; vol. 1952; 2018. pp. 1-9

[19] Moises TT, Enrique MG, Antonio OT. An approach of efficient security algorithms for distribute architectures. In: 2017 International Conference on Energy, Communication, Data Analytics and Soft Computing, ICECDS 2017; 2018. pp. 22-25

[20] Toapanta SMT, Anchundia MAM, Mafia LEG, Orizaga JAT. Biometric systems approach applied to a conceptual model to mitigate the integrity of the information. In: CITS 2018-2018 International Conference on Computer, Information and Telecommunication Systems; 2018. pp. 1-5

[21] Toapanta SMT, Cruz AAC, Gallegos LEM, Trejo JAO. Algorithms for efficient biometric systems to mitigate the integrity of a distributed database. In: CITS 2018-2018 International Conference on Computer, Information and Telecommunication Systems; 2018. pp. 1-5

[22] Toapanta SMT, Gallegos LEM, Quimi FGM, Trejo JAO. Analysis of efficient processes for optimization in a distributed database. In: CITS 2018-2018 International Conference on Computer, Information and Telecommunication Systems; 2018. pp. 1-4

[23] Toapanta SMT, Paredes SJM, Gallegos LEM, Trejo JAO. Analysis of HIPAA for adopt in the information security in the civil registry of the Ecuador. In: CITS 2018-2018 International Conference on Computer, Information and Telecommunication Systems; 2018. pp. 1-5

[24] Toapanta M, Mero J, Huilcapi D, Tandazo M, Orizaga A, Mafla E. A blockchain approach to mitigate information security in a public organization for Ecuador. In: IOP Conf. Ser. Mater. Sci. Eng. 2018;**423**(1):1-7

[25] Toapanta M, Nazareno J, Tingo R, Mendoza F, Orizaga A, Mafla E. Analysis of the appropriate security models to apply in a distributed architecture. In: IOP Conf. Ser. Mater. Sci. Eng. 2018;**423**(1):1-5

[26] Mafla E, Toapanta M, Barona D, Contrera G. Optimization of an electronic signature scheme in a voting system in a distributed architecture. In: 2018 3rd International Conference on Computer Science and Information Engineering (ICCSIE 2018); 2018. pp. 242-248

[27] Mafla E, Toapanta M, Huilcapi D, Cepeda M. Ensuring the blind signature for the electoral system in a distributed environment. In: 2018 3rd International Conference on Computer Science and

Information Engineering (ICCSIE 2018); 2018. pp. 205-212

[28] Mafla E, Toapanta M, Orellana N, Barona D. Analysis cryptographic for electronic votes in systems of distributed architectures. In: 2018 3rd International Conference on Computer Science and Information Engineering (ICCSIE 2018); 2018. pp. 189-196

[29] Mafla E, Toapanta M, Tamayo J, Ortiz J. An approach to the efficient security algorithms used in voting scanning in an electoral process. In: 2018 3rd International Conference on Computer Science and Information Engineering (ICCSIE 2018); 2018. pp. 234-241

[30] Mafla E, Toapanta M, Chávez L, Ortiz J. A homomorphic encryption approach in a voting system in a distributed architecture. In: 2018 3rd International Conference on Computer Science and Information Engineering (ICCSIE 2018); 2018. pp. 182-188

[31] Maciel R, Toapanta M, Mendoza F, Plùa D, Tandazo M, Mafla E. Analysis of security algorithms for a distributed database. In: 2018 3rd International Conference on Computer Science and Information Engineering (ICCSIE 2018); 2018. pp. 197-204

[32] Mafla E, Toapanta M, Espinoza J. A hyperledger scheme for the deployment of smart contracts in a public organization of Ecuador. Smart Innovation, Systems and Technologies. 2018;**1**:1-13

[33] Mafla E, Toapanta M, Aguilar J. Analysis of adequate bandwidths to guarantee an electoral process in Ecuador. Smart Innovation, Systems and Technologies. 2018;**1**:1-12

[34] Mafla E, Toapanta M, Coronel M. Appropriate security protocols to mitigate the risks in electronic money management. Smart Innovation, Systems and Technologies. 2018;**1**:1-12

[35] Mafla E, Toapanta M, Mendoza F. Cryptographic algorithms to mitigate the risks of database in the management of a smart city. Smart Innovation, Systems and Technologies. 2018;**1**:1-12

[36] Mafla E, Toapanta M, Mendoza F. Impact on the information security management due to the use of social networks in a public organization in Ecuador. Smart Innovation, Systems and Technologies. 2018;**1**:1-13

[37] Joseph Guaman AO, Toapanta M. An information security approach in the Armed Forces of Ecuador. In: CITIS 2017; 2017. pp. 366-372

[38] Mafla E, Toapanta M, Orizaga A. Algoritmos y protocolos de seguridad para el registro civil del ecuador. Universidad de Guadalajara. Tesis doctorado, Identificador de la entrega; 2018:930484533

[39] Dennis KH. On the use of the Clark-Wilson security model to protect industrial automation control systems. In: CSIIRW'13 Proceedings of the Eighth Annual Cyber Security and Information Intelligence Research Workshop; 2013

[40] Hermann E. The limes security model for information flow control. In: A. Dept. of Secure Inf. Syst., Upper Austria Univ. of Appl. Sci., Linz; 2011

[41] Cui J, Wang Q. An improved BLP model with response blind area eliminated

[42] Manr D, Servicio WMT, Civil R. Seguridad de la Información. Facultad de Ingenieria Universidad de Buenos Aires; 2005. Available from: http://www.cicomra.org.ar/cicomra2/expocomm/TUTORIAL5Heguiabehere-CERTANT.pdf [Accessed: March 19 2018]

[43] Cucurull J, Puiggali J. Distributed immutabilization of secure logs. In: Security and Trust Management, STM 2016; vol. 9871. 2016. pp. 122-137

[44] Parkavi R, Babu KRC, Kumar JA. Multimodal biometrics for user authentication. 2017. pp. 501-505

[45] Ogiela MR, Majcher M. Security of distributed ledger solutions based on blockchain technologies. In: Proceedings-International Conference on Advanced Information Networking and Applications, AINA; vol. 2018: no. c; 2018. pp. 1089-1095

[46] Sukhwani H. Performance modeling of hyperledger fabric (permissioned blockchain network). In: 2018 IEEE 17th International Symposium on Network Computing and Applications (NCA); 2018. pp. 1-8

[47] B. with S. using a theme provided by R. the Docs., "Hyperledger Fabric," Writing Your First Application (tutorials, 2019. [Online]. Available: https://hyperledger-fabric.readthedocs.io/en/release-1.4/write_first_app.html. [Accessed: April 11 2019]

# INDEX

9 789390 013852

Printed at Repro India Ltd.